T0326461

E-Learning and Education for Sustainability

Environmental Education, Communication and Sustainability

edited by Walter Leal Filho

Vol. 35

Zu Qualitätssicherung und Peer Review der vorliegenden Publikation

Die Qualität der in dieser Reihe erscheinenden Arbeiten wird vor der Publikation durch externe, vom Herausgeber benannte Gutachter im Double Blind Verfahren geprüft. Dabei ist der Autor der Arbeit den Gutachtern während der Prüfung namentlich nicht bekannt; die Gutachter bleiben anonym.

Notes on the quality assurance and peer review of this publication

Prior to publication, the quality of the works published in this series is double blind reviewed by external referees appointed by the editor. The referees are not aware of the author's name when performing the review; the referees' names are not disclosed.

Ulisses Miranda Azeiteiro / Walter Leal Filho / Sandra Caeiro (eds.)

E-Learning and Education for Sustainability

Bibliographic Information published by the Deutsche Nationalbibliothek
The Deutsche Nationalbibliothek lists this publication
in the Deutsche Nationalbibliografie; detailed bibliographic
data is available in the internet at http://dnb.d-nb.de.

Library of Congress Cataloging-in-Publication Data
E-learning and education for sustainability / Ulisses Miranda Azeiteiro, Walter
Leal Filho, Sandra Caeiro, eds.
    pages cm. -- (Umweltbildung, Umweltkommunikation und Nachhaltigkeit
= Environmental education, communication and sustainability, ISSN
1434-3819 ; Vol. 35)
 Includes index.
 ISBN 978-3-631-62693-1 -- ISBN (invalid) 978-3-653-02460-9 (E-Book) 1.
Environmental education. 2. Sustainable development--Study and teaching.
3. Distance education. I. Azeiteiro, Ulisses, editor of compilation. II. Leal
Filho, Walter, editor of compilation. III. Caeiro, Sandra, 1977-
 GE70.E14 2014
 333.7071--dc23

                          2014024488

                     Copy editing and layout:
                     Kumpernatz + Bromann
                   www.kumpernatz-bromann.de

                       ISSN 1434-3819
                  ISBN 978-3-631-62693-1 (Print)
                E-ISBN 978-3-653-02460-9 (E-Book)
                  DOI 10.3726/978-3-653-02460-9

Table of Contents

Editorial

This book aims to contribute to the global debate on the implementation of education for sustainability, by discussing the use of e-learning in contributing towards sustainable development as a whole, and education for sustainable development in particular. Also it aims at reiterating the role that e-learning can play in this process, linking pedagogical concepts with curricular issues, at the same time that it illustrates the wide range of technological possibilities available, which may help to achieve the intended the sustainability-oriented learning outcomes.

Almost three decades after the concept of sustainable development was formally put in the international agenda by the Brundtland Commission, the year 2014 is a year where historical goals should have been reached, since it is the last year of the United Nations Decade on Education for Sustainable Development (UNDESD). Within this decade, research, projects and educational initiatives were undertaken and various deliverables were produced. E-learning is one of the areas where some developments have been seen, since it has become more widely accepted in formal and non-formal education settings, with a proven potential to be an effective tool towards promoting education for sustainable development.

The use of Information and Communication Technologies (ICT) as a whole, and of e-learning in particular, presents many advantages. Finally, these technologies cater for much content and for various means of delivery. Secondly, the delivery of content is flexible and can be adapted to various circumstances and settings. Furthermore, the use of the internet and social networks have helped to make e-learning on education for sustainable development more popular, and also more present in learning and education processes. Finally, the time and space flexibility associated with e-learning contributes even more for its growing use.

Bearing in mind the broad field of e-learning and sustainability, this book is divided in four main parts: The first part is more related with principles, concepts and competences. The second part focuses on ICT tools, materials and teachers' skills. The third part presents examples of good practices in emerging countries in Africa and Asia, whereas the final concentrates on formal and non-formal teaching and learning experiences.

A total of 18 double-blind peer-reviewed papers from Europe (11), Australia (2), Asia (1), North-America (2) and Africa (2), cover the different subjects related to the above themes of this book.

In the first part of the book related with principles, concepts and competences readers can found four chapters:

A first chapter from Anne Sibbel entitled "An experience in developing and implementing blended learning for sustainability". This chapter is about implementing education for sustainability, and it is suggested that higher education curricula can be adapted to respond to the challenge of sustainability through a collaborative and reflective approach to teaching and learning.

Joop de Kraker, Ron Cörvers and Angelique Lansu in he chapter "E-learning for sustainable development: linking virtual mobility and transboundary competence development" review how competences for sustainable development have been defined for higher education, and argue that a key competence to be developed is 'transboundary competence' (that authors define it as the ability for productive interaction across the boundaries between different perspectives). The authors apply the principles of competence-based learning, to derive the design principles for effective learning environments to foster transboundary competence.

In a globalisation context, environmental professionals have to develop social, ethical, creative, personal and interpersonal skills in addition to technical competences to be of value in attaining sustainability. These skills are also necessary for university environmental graduates to enter the labour market, and improve their employability. In the chapter "Training and employability, competences from an e-learning undergraduate programme in environmental sciences", Ana Paula Martinho, Sandra Caeiro, Fernando Caetano, Ulisses Miranda Azeiteiro and Paula Bacelar-Nicolau assess the development and acquisition of key skills and competences in the 1st cycle degree programme of Environmental Sciences at the Universidade Aberta, the Portuguese Distance Learning University, and their contribution to the employability of its graduates discussing the results within the European framework for higher education programmes.

Within sustainability issues, climate change is recognised as one of the most challenging and defining themes for our future. In the fourth chapter, titled "Transforming academic knowledge and the concept of lived experience: intervention competence in an international e-learning programme" Francisca Pérez Salgado, Gordon Wilson and Marcel van der Klink, consider the concept of the Lived Experience to be important and perhaps even crucial for the domain of sustainability, where it can be used to expand knowledge and linking academia with professionals and citizens. For the authors the concept of the Lived Experience explains the existence of several perspectives at the same time, connects abstract and distant scientific knowledge with personal, local and cultural diversity and it considers epistemological diversity as a resource for social learning and holistic knowledge.

The second part of the book about ICT tools, materials and teachers skills is divided in seven more chapters (chapter 5 to 11):

Daniel Otto in "Let's Play! Using simulation games as a sustainable way to enhance students' motivation and collaboration in Open and Distance Learning" introduces simulation games within an Open and Distance Learning (ODL) context. Recent developments in technology have bridged the gap of adequate tools which have long been identified as the central hurdle in utilizing simulation games for ODL. Based on its characteristic features, simulation games seem particularly suitable for learning about environmental topics.

In the chapter "Developing e-Learning materials for teaching industrial ecology and environmental sustainability", Anthony Halog and Gary Dishman state that the urgent call for climate change mitigation and the need in the pursuit of the vision of sustainable, circular economy, an increasing number of universities have been developing courses to prepare students to meet the growing demand for green and sustainability related jobs, which require systems perspective training in Industrial Ecology (IE) and Life Cycle Assessment (LCA). This chapter contributes to develop an online-based information system, which provides an interdisciplinary approach to teach IE and LCA for sustainable industrial development.

Sally Caird, Andy Lane and Ed Swithenby in the chapter "Greening Higher Education qualification programmes with online learning" introducesthe SusTEACH Modelling Tool, developed following a research on the impact of computing technologies or Information and Communication Technologies (ICTs) on HE teaching models, together with a carbon-based environmental assessment of 30 courses, offered by 15 UK-based HE institutions. Their discussion includes the role of online learning designs and pedagogical use of computer technologies for achieving carbon reduction in HE.

Joop de Kraker and Ron Cörvers in their chapter entitled "European Virtual Seminar on Sustainable Development: international, multi-disciplinary learning in an online social network" introduce readers to the European Virtual Seminar on Sustainable Development (EVS), which is a web-based course that aims to foster competences for sustainable development through collaborative learning in virtual, international, multi-disciplinary student teams. The chapter focuses on the recent adoption of a social networking platform to enhance the sociability of the EVS virtual learning environment. The new learning environment is evaluated in terms of student experiences and perceptions, actual tool use, and team performance.

In the chapter "Electronic logistics for a sustainable distance education: the new UNED on-site virtualization of evaluation procedure documents" by Mari Carmen Ortega-Navas, Rocío Muñoz-Mansilla, Fernando Latorre and Rosa María Martín-Aranda, the authors illustrate how the Spanish National Distance Education University (UNED) has implemented a new protocol for evaluation procedures, that optimizes paper use and transportation through on-site digitalization of exams, bringing a new logistics paradigm. UNED evaluation system simultaneously summons students at many locations in Spain and at selected venues across the world. Technology is the main contribution in the so-called "valija virtual" (virtual attaché case) system, which has been developed at UNED.

Leanna Archambault and Annie Warren in the chapter "Leveraging E-learning to prepare future educators to teach sustainability topics" describe Sustainability Science for Teachers, a hybrid course in development at Arizona State University that integrates the use of technology and digital storytelling to teach sustainability topics in a meaningful way. This course is required as part of a programmatic education reform aimed at improving science content knowledge among pre-service teachers. The goal of the course is for future educators to gain necessary knowledge and skills about sustainability, allowing them to become more informed citizens and helping them learn how to address sustainability concepts within their future classrooms.

In this second part, the last chapter "The use of information and communication technologies by the secondary school teachers for developing a more sustainable pedagogy in Latvia" from Dzintra Ilisko and Svetlana Ignatjeva presents the analysis of the use of information and communication technologies in the secondary school setting, as well as the sustainability of the purpose of the use of the ICT. The authors discuss the pedagogy that is equitable with the use of the ICT technologies. A special emphasis is given to the problems and barriers which prevent secondary school teachers to integrate ICT in their teaching, as well as the main purpose why teachers are willing to integrate technologies in their teaching. Finally, the authors outline suggestions for teacher trainers on how to integrate ICT in developing more sustainable teaching and on how teachers can improve their ICT literacy to improve their teaching skills. The article poses questions on whether the current pedagogical models are compatible with ICT, and if the integration of ICT can improve learning. Teachers who are trying to improve their teaching are facing new challenges of how to adopt new approaches and how to integrate new technologies in their teaching.

In the third part of this book, readers can find three examples of good practices from emerging and developing nations (chapters 12 to 14).

In the chapter "A critical narrative of e-learning spaces for sustainable development in the Global South" Rudi W. Pretorius says that the implementation of e-learning poses several challenges for teaching staff, students and administrators in contexts such as the Global South – a collective name for emerging and developing nations. Although these challenges vary from country to country, limitations in terms of access to the Internet – even among university students – remains a barrier for the effective roll-out of education to exactly those segments of the community needing it the most. Since Education for Sustainability per definition cannot exclude certain community segments, implementation of e-learning in contexts as the Global South should proceed with due consideration of matters such as the latter.

In "Cotonou 2012 and Beyond – An assessment of e-learning for sustainability in sub-Sahara Africa" J. Manyitabot Takang and Christine N. Bukania assess e-learning for sustainability in sub-Sahara Africa, by drawing on some case studies such as the African Virtual University. The assessment investigates to what extent the e-learning initiatives are home-grown, i.e. specifically designed to solve local problems, how e-learning is bridging the lack of human capacity, and access to information on sustainability. Moreover, this chapter addresses the question of how inclusive e-learning for sustainability could be in Africa, especially in view of infrastructural limitations and the costs of participating in e-learning courses. The authors conclude with some recommendations for future developments in e-learning for sustainability in Africa.

Prakash Rao, Yogesh Pati, Manisha Ketkar, Viraja Bhat and Shilpa Kulkarni in the chapter "Sustainability in an educational institution: analysing the transition to paperless e-processes, an Indian case" focus on a new approach towards a sustainability based academic process, through the development of an online system using Information Technology tools at an Indian Business school. This includes tools like Faculty Information Systems, online testing, online student performance appraisal and feedback systems, among others, which have been designed to create improved academic efficiency, higher stakeholder satisfaction levels along with potential environmental co benefits like reduced paper usage. The need for incorporating sustainability as a core element of higher education has been growing over the years in Indian universities, and the study is a step towards providing solutions by using IT systems. Key lessons learnt from the study revealed that replicability of this approach in other institutions is dependent on several factors like stakeholder satisfaction, infrastructure, skilled workforce, top management support and process standardisation.

The fourth and last part of this book is about formal and non-formal successful experiences.

Deep learning is seen as a way to maximize the benefits from sustainability education for both students and society, and Amelia Clarke in the chapter "Building an online Master's Program for Deep Learning in Sustainability" describes the University of Waterloo's Master of Environment and Business (MEB) programme in relation to seven characteristics of deep learning. The MEB program is an executive education programme that is mostly delivered online. Student survey responses show a high satisfaction with the programme, and highlight areas where deep learning is occurring. This chapter emphasizes that it possible to ensure deep learning in an online program and in a program's design (not just at the teaching activity/ assignment level). The chapter ends by offering lessons learned for other online courses and programmes.

In the chapter "First online course on desalination by renewable energies, lessons learnt", Juan A. de la Fuente, Vicente J. Subiela and Baltasar Peñate describe a training initiative that was developed by the ITC under the framework of the PRODES project which was co-financed by the European Commission within the "Intelligent Energy for Europe Programme". The e-learning course called "Introduction to desalination by renewable energies" is addressed to those people who could be interested in this field of knowledge, as professionals related with water or energy sectors, or technology students. The e-learning course on desalination by renewable energies sets the fundamentals on using renewable energy for desalination, as a contribution to address the world's water situation by increasing the knowledge on these technologies. After several editions made the international impact is very relevant. After this experience, the ITC is committed to improve the course in each edition; an extended and upgraded version of the course has already concluded with the collaboration of the European Desalination Society (EDS).

Luísa Aires, Paulo Dias, José Azevedo, M. Ángeles Rebollo and Rafael García-Pérez in the chapter "Education, Digital Inclusion and Sustainable Online Communities" propose a theoretical framework to analyze the relations between online communities, participation and digital inclusion. Sustainability education is interpreted as a shared and interactive process and is associated to community development across collaborative practices. In this framework, sustainable online communities embody the change of educational cultures. The authors state that over the last decade, significant contributions have been made to a better understanding of sustainability and education. However, very little has been researched on sustainability education in the digital society. In new digital environments, collaborative practices, hybrid contexts, distributed knowledge or shared responsibility give rise to new approaches to the relationship between education, participation and digital inclusion – central constructs of sustainable education. The authors conclude that connecting digital inclusion and empowering practices is a way to develop sustainable online education.

A final contribution is provided by Walter Leal Filho, who briefly outlines some of action needed in order to foster the use of e-learning on education for sustainable development.

Given the variety of research, this book offers a diverse thematic and geographic overview of some current of the key issues and good practices in E-Learning and education for sustainability. In addition, the chapters address some important challenges and future developments, also giving insights into how education for sustainability through e-learning may be pursued, despite different subjects and educational contexts.

We would like to take this opportunity to thank all authors who submitted their manuscripts for consideration of inclusion in this book. And since the peer review was a double-blind process, we also thank the reviewers who have taken time to provide timely feedback to the authors, thereby helping the authors to improve their manuscripts, and ultimately the quality of this book.

Ulisses Miranda Azeiteiro, Walter Leal Filho and Sandra Caeiro

I.
Principles, concepts and competences

An experience in developing and implementing blended learning for sustainability

Anne Sibbel[1]

Abstract

Implementing education for sustainability involves questioning accepted practice, theory and values by both teachers and their students. Based on principles of knowledge management, and concepts pioneered by some early, liberal educational theorists, some curriculum innovations were implemented to develop university students' capabilities for sustainability. An online hub was established for knowledge collection, guided analysis and re-construction by the students and teacher. The challenges presented via by a series of online tasks were varied, but all were scaffolded with links to pertinent information, and with practice in specific learning experiences offered in the immediately preceding face-to-face class. Some tasks involved interpretation of key documents outlining relevant international or national policies or positions of government and other influential stakeholders. Other tasks required self-assessment of the sustainability of personal behaviours, and then assessment of real-life small scale and large scale operations, after analysis of primary or secondary data. For each task, feedback related directly to evidence of developing capabilities for sustainability was provided confidentially to each student. A synthesis and evaluation of all responses was also presented online, highlighting anonymous examples of personal insights, ideas and values aligning to sustainability. In this way, accumulating knowledge was shared, analysed and re-constructed for individual and collective reflection by students, influencing the ways they approached the next task. Outcomes suggest that higher education course curricula can be adapted to respond to the challenge of sustainability through this collaborative and reflective approach to teaching and learning.

Introduction

The responsibilities of universities for preparing graduates to be leaders in work towards sustainability are well described and widely accepted (Clark and Button, 2011; Ferrer-Balas et al., 2010; Hesselbarth and Schaltegger, 2014). Education for sustainability involves raising awareness and developing students' understanding and motivation to critically reflect on day-to-day behaviours, operations and practices, in terms of their impact on society and the environment. This presents challenges to conventions in pedagogy in higher education. Efforts in changing to education for sustainability inevitably compete with the many 'educational reforms towards efficiency, accountability, privatization, management and control' (Wals, 2014, p. 14). In part, this may explain why much of the curriculum development in this area has been driven by the interests of a few so that efforts have been 'piecemeal and discipline-based' (Stubbs and Schapper, 2011, p. 260).

---

1   RMIT University, Melbourne, Australia.

Change is ongoing in the higher education sector. At the curriculum level, new information is added to course content to ensure currency and relevance, and technologies offer new ways to teach and learn. Despite this dynamic milieu, most changes tend to take place without challenging traditions (Miller et al., 2011). More radical innovation in teaching and research is required if universities are to become change agents for sustainability (Peer and Stoeglehner, 2013).

It is possible to describe many approaches to integrating sustainability into university curricula (Lozano and Young, 2013; O'Brien and Sarkis, 2014), ranging from the addition of relevant content into an existing course to developing whole programs or specialisations which are dedicated to education for sustainability. Somewhere between these approaches, sustainability and related concepts might be applied in outreach and campus activities, or embedded into a stand-alone sustainability course offered across a range of disciplines, or into a number of courses offered within a single discipline. Through strategically targeting courses with large enrolments, or through professional development for staff who teach many classes, the impact of some of these initiatives can be enhanced. Despite some successes even across whole institutions (for example Adomssent et al., 2007; Beringer, 2007; Holmberg et al., 2012), there is a long way to go before education for sustainability is universal in this sector.

Managing knowledge for sustainability

It is through its roles in producing new knowledge, and in exploiting that knowledge in new applications, that higher education exerts it influence on society (Adomssent, 2013). Universities have responsibilities for brokering the 'two way learning and flow of knowledge' (Dlouha et al., 2013a, p. 64), a process which involves networking and relationship building with society (Sheate and Partidario, 2010). Decision-making for sustainability requires knowledge derived through processes within social systems, rather than simply knowledge sharing and transfer (Miller et al., 2011). While universities have a 'unique role in deepening and expanding human knowledge … it is precisely a lack of knowledge integration and pertinent use of that knowledge, which is at the root of the current crises' (Ferrer-Balas et al., 2010, p. 607). The need to address the 'knowledge gap' and the 'under provision of knowledge' in many sectors of society (Adomssent, 2013, p. 13) further highlights the important relationship between sustainability and the ways that knowledge is managed by universities to build social capital.

Sustainability knowledge has some unique characteristics. It is 'socially robust', relies on a range of epistemologies rather than a singular disciplinary perspective, recognises complexity in systems rather than controlling or eliminating variables, and is subject to values and ethical considerations (Miller et al., 2011, p. 179). It differs from knowledge typically presented in academic studies which is organised within single traditional disciplines. The increasingly dynamic, accessible and pervasive information environment available through digital technologies has become an important resource in contemporary education. Nonetheless, information must be 'connected, organized and relevant' (Lopez, 2013, p. 294), if learners are to understand and respond to the challenges of sustainability. New ways of engaging with and organising information are required for transformative learning. Knowledge management principles can be applied to guide collection, exchange, integration and evaluation to construct new knowledge to bring about change. These principles offer a useful framework for reflecting on the requirements of a curriculum, as a system of organising and utilising resources to promote learning for sustainability.

Pedagogical frameworks

Education for sustainability is transformative education. Transformative learning involves collaborating with others to challenge existing worldviews and construct new knowledge for dealing with unsustainable conditions (Cebrian et al., 2013). Much more than the acquisition and reproduction of knowledge (Posch and Steiner, 2006), education for sustainability must generate motivation and inspire commitment to work towards sustainability. The changes in values and attitudes, which will encourage graduates to take on this challenge and to lead others to do so, involve the affective domain (Shephard, 2008). This contrasts directly with transmissive educational approaches, which focus on the cognitive domain, so tend to reinforce a set of values embedded in past practice.

According to Bloom's taxonomy of educational objectives, learning involves progressing through sequential levels of cognitive difficulty, from information gathering and comprehension, through to application and analysis, then synthesis and evaluation of the knowledge. While Bloom's approach has been criticised for its emphasis on cognitive development, it provides a useful platform on which to construct a curriculum model designed to achieve affective learning outcomes. In fact, some recent research (Pappas et al., 2013; Tello et al., 2013) has confirmed the relevance of Bloom's taxonomy of educational objectives to education for sustainability. Many of the constructivist models of learning used to guide contemporary curriculum design align with Bloom's hierarchy because they assume that higher order cognitive skills build on lower order skills. But these models also offer the scope to integrate features, which facilitate collaborative learning and encourage self-reflection, as required for transformative learning. Within the scope of this paradigm, theories of social constructivism recognise the significance of social and cultural influences on learning. For instance, Vygotsky's Sociocultural Theory (Miller, 2011), which emerged early in the 20th Century, describes how individuals learn through interactions with others who provide support with encouragement, modelling and suitable resources. Building on this work, the constructivist Bruner introduced the concept of scaffolding to refer to the support offered to a learner to complete a task which could not be achieved otherwise (Wood et al., 2006). Learners can be scaffolded to understand content, or to engage in the social interactions which lead to learning (Salmons, 2009), and include technology-based supports within e-learning environments (Yu, 2009). It is the scaffolding due to interactions in a social context which can lead to higher order thinking and metacognition (Macgregor and Turner, 2009). The relevance of a range of constructivist pedagogies to understanding learning through social processes, including those provided in e-learning environments, has been recognised by others (see for example, Barth and Burandt, 2013; Macgregor and Turner, 2009).

Drawing from all these ideas, it seems that a setting with open, inviting and scaffolded opportunities for participation and collaboration is a critical requirement for transformative learning. This informed the approach to a curriculum for sustainability described here.

Online learning technologies

Technology has had a major influence on the ways that information is generated and accessed, and increasingly, on teaching and learning in higher education (Dlouha et al., 2013b). A growing body of research has described the benefits of online technologies in this context. For instance, online learning resources can be readily updated or adapted over time to ensure currency and relevance (Macgregor and Turner, 2009). Online technologies can offer a network of accessible communication pathways with many interfaces for potential collaboration, necessary for construction of new knowledge. Some of the requirements for best practice in learner-centred environments include 'frequent and ongoing feedback' to students and 'frequent student performances' (Habron et al., 2012, p. 381). Technologies offer many different tools to facilitate and encourage these patterns of regular participation and assessment.

The structure of online networks is an important variable influencing whether participation leads to transformative learning. Inclusive and open networks, distributing control through non-hierarchical structures, are consistent with enabling participants to be both producers and consumers of knowledge (de Kraker et al., 2013). Another feature often observed on informal social networking sites is the merging of emotional, social, cultural and professional perspectives. All these features are consistent with a participative and whole person approach to learning, as required in education for sustainability.

A number of limitations of online learning systems have also been identified, including the potential for information overload (Sridharan et al., 2010). Despite the accessibility of an online network, there are many reasons why students may be reluctant to participate. Generally, the competitive academic environment may discourage students from sharing their information sources and ideas with their peers in the real or virtual classroom. Dissonance between institutional practices and sustainability-related learning objectives can also discourage student involvement (Barth, 2013). Participation in online environments is associated with choice based on the perceived value of accessing available knowledge, of the socialisation or other opportunities offered there (Seraj, 2012). University students may find that the obvious functional similarities between academic online learning sites and social networking sites create a 'role conflict' which could discourage participation in formal online learning activities (de Kraker et al., 2013, p. 121). There are other features of informal online social networks which are not necessarily conducive to collaboration leading to learning for sustainability. For instance, the idea of openly rating others' contributions could discourage or otherwise restrict participation, so runs contrary to a supportive and inclusive network. Such limitations need to be considered when relying on e-learning strategies as part of a curriculum for sustainability.

The 'blended' learning approach involves the combination of web-based and face-to-face strategies for teaching and learning (Mahmood and Hafeez, 2013). This broad definition accommodates many different conditions for learning, from individual to collaborative, from teacher-directed to student-led activities, from the classroom to any other setting. There is no universally accepted definition, other than the substitution of a proportion of traditional face-to-face instruction with online learning opportunities (Graham et al., 2013; Owston et al., 2013). The components can be blended in 'virtually unlimited possible combinations' (Moskal et al., 2013, p. 15), in ways described as 'thoughtful' or 'effective' (Garrison and Kanuka, 2004, p. 96, 97), through to more specific proportions of course content offered in virtual relative to real classroom environments (Ross, 2012). However, the amounts of time or activities allocated to each environment do not necessarily determine the effectiveness of blended learning. It depends on how the 'affordances' of online technologies and conventional teaching are integrated, and reflected in the learning experiences (Graham and Dziuban, 2008).

Others have noted that e-learning curricula should be guided by pedagogy, rather than led by technological possibilities, if they are to deliver the intended learning outcomes (Csete and Evans, 2013). A review of research undertaken in the last few years has demonstrated that a blended approach can lead to higher levels of student satisfaction and achievement, and can reinvigorate teaching practice (Owston, 2013).

Much of this research has focused on the influence of practical aspects of delivery, at the expense of building a theoretical basis to guide practice (Drysdale et al., 2013; Halverson et al., 2014). To make a contribution to this emerging field, this project explored the possibilities for transformative learning of a blended curriculum model based on theories of knowledge management and some well-accepted aspects of constructivist pedagogy.

The curriculum development project

This project involved a single course designed to teach principles, applications and values aligned with sustainability to undergraduate students studying second or third year sciences, including social sciences. Since its inception in 2000, the course was taken as an elective, and for many, may have been the only opportunity to engage in formal education for sustainability before they graduate. Education for sustainability requires that students reflect on the epistemology and discourse of their own disciplines, and then consider how they could bring about change in the fields of their future professional practice. Given the powerful influence of established knowledge systems within higher education, enabling students from many different disciplines to meet this objective through a single elective course presented a formidable challenge.

The aim of this project was to explore the potential of a combination of face-to-face and online interactions, guided by a constructivist pedagogical approach and principles of knowledge management, to bring about affective learning outcomes. Following these principles, attention was focused on the processes of capture, interpretation, integration and re-construction of knowledge through student and teacher involvement online. These processes were supported by face-to-face interactions to develop the potential of a course to promote transformative learning. Rather than assessing actual learning, the focus of this paper is on describing how the pedagogy and knowledge management principles were applied in designing and implementing the curriculum. As an exploratory study, evaluation was formative. These formative evaluations are outlined in general terms, so that they may be useful to others planning to adapt their teaching towards education for sustainability in different contexts.

The project context

For more than a decade, the course had been offered once each year to relatively small classes of up to 45 students. This period was a particularly formative time, set against a backdrop of emerging understandings of the concept and applications of sustainability, as well as the changing perceptions of the roles and responsibilities of universities. There were other dominant influences on the curriculum for this course.

Coinciding with changing expectations of students about university learning environments was the availability of more reliable and sophisticated online technologies to support their learning. While this was leading to greater flexibility in teaching across the university, considerable attention was directed to academic accountability for delivering prescribed content and on ensuring rigorous assessment in this new learning environment. Overlaying all these variables were the growing public awareness about sustainability related issues, and the accumulating evidence establishing an imperative for changing behaviours in personal and professional lives.

The course

The curriculum had developed through many course iterations by 6 academic staff as well as several other professionals with leadership roles in various areas of sustainability. New expertise, experience and knowledge had been injected into the course each time it was offered. After each iteration, summative learning outcomes guided the review and revisions to pedagogy, content and assessment tasks. The legacy of this history was a rich compilation of learning resources, teaching and assessment strategies.

Several topics were consistently covered by the course each year. In the first sessions, students were introduced to aspects of the history of the development of the concept of sustainability, its contested nature and scope, and then its application to an increasingly wider range of problems and contexts as the course progressed. Emphasis was always placed on considering the economic, environmental and social dimensions of any problems, whether through analysis of case studies or in projects selected by students. This analysis was contingent on an awareness of the tools used for sustainability assessment, another important topic covered in this course. Finally, strategies for individual and collective action for sustainability, including educating others for sustainability, were devised and evaluated. Over this time, self-reflection on personal cultural frameworks, values and attitudes increasingly became recognised as part of the learning process.

The course had always offered a blended approach to learning, and as new e-learning technologies became available, the proportion delivered online increased. In 2012, online materials, tasks and opportunities for interactions were presented on alternate weeks to the face-to-face lectures, in-class discussions and activities. Even more than in previous years, the intention was to develop students' autonomy in directing their own learning. So while many resources were accessible on the course website, the tasks progressively required students to source their own materials, then to make and share judgements about their validity, reliability and value before use. As recommended by Macgregor and Turner (2009), teaching the competencies required to make these judgements was embedded in the course. This was one of the ways that the balance was further adjusted away from teacher-guided to student-led learning pathways.

Assessing learning outcomes

Delineating or defining learning outcomes is always problematic in education for sustainability. It relies on identifying the capabilities for sustainability, in terms of competencies, skills, values and attitudes. These capabilities have been postulated and explored in an extensive body of literature over many years (see for example Barth et al., 2007; de Haan, 2006; Posch and Steiner, 2006; Wiek et al., 2011; Wiek et al., 2013).

Although there is no international consensus on these capabilities (Rieckmann, 2012), there have been many attempts to describe, discriminate, categorise or to embrace them in concepts such as 'action competence' or 'Gestaltungskompetenz' (Mochizuki and Fadeeva, 2010; de Haan, 2006). Generally they tend to refer to learner autonomy and self-efficacy, as well as the political skills of communication, negotiation, advocacy and conflict resolution. Students also need to develop an awareness of their social and moral responsibilities, with values aligned to sustainability so they are willing to work beyond self-interests. As well as balanced and current knowledge of global issues, they need experience in dealing with real world problems.

The learning objectives for the course had been based on these wide ranging capabilities and were subject to review each year as new insights into the challenge of sustainability emerged.

Equitable and accountable assessment relies on shared understandings of a set of criteria, which can be readily applied to make valid and objective measurements of individual and whole class learning. For this curriculum, assessment criteria had to relate to transformative learning outcomes. Imposing precise learning targets could counter the goals of transformative learning which is necessarily open-ended and student-led. At the same time, more open-ended or abstract criteria might frustrate or confuse students. It was necessary to strike a balance here by presenting each task with the respective assessment criteria, then adjusting them after negotiation with the students. Further revisions to the criteria for subsequent assessments were guided by the results of each task and student feedback. Along with the other important curriculum components, these criteria were being developed through cycles of participation and on-task activity, reflection and change.

Assessment Tasks

A series of assessment tasks corresponding to the learning objectives was developed. To respond to each task, the students were required to present information they found, then organize it in meaningful ways to explain their ideas to each other. In this preliminary phase of the knowledge management cycle, learning from peers was the means through which students could develop their own knowledge. Once all contributions were captured online, anonymous examples of insights into a problem, as well as innovative and resourceful ideas for resolution could be identified. They were organised and evaluated by the teacher as a whole class response to the challenges presented by a task. With students from a range of disciplines participating, this synthesised class response would not be constrained by discipline-centric thinking. In the process leading to the construction of this new knowledge, the teacher raised questions, exposed gaps in existing knowledge, and offered alternative perspectives.

Feedback was also provided to each student confidentially, highlighting evidence of their emerging capabilities for sustainability, areas where assumptions could be checked, and with relevant resources to guide further reflection. It was anticipated that this confidential feedback, combined with the opportunity to compare all the individual responses and the whole class response displayed online, would influence how students approached the next task. Overall, this cycle of knowledge management, that is collection and sharing, then integration and reconstruction, was designed to encourage self-reflection to bring about change in personal attitudes and values.

Measuring ecological footprints has been recognised as a way to raise individuals' awareness about the relative impact of their lifestyles. In making human impact more 'tangible' (Savageau, 2013, p. 17), such assessments are more likely to trigger reflection and review of personal behaviours. Using some publicly accessible software, students were asked to undertake a personal audit of their usual patterns of energy and resource consumption. They were provided with some sample audits and simple calculation tools to support their learning and enable them to make the necessary estimates. Students posted their individual results online, including the effects of modifying some of their behaviours to reduce impact. Sharing personal reflections on the feasibility of these changes was the first stimulus for group interaction in class, with the teacher modelling an encouraging and non-judgemental approach to all contributions. This planned transition from individual to group work was an initial step towards advancing the idea of collaboration, as a key to innovative problem solving for sustainability.

A problem based approach can optimise opportunities for learning (Habron et al., 2012). Students were asked to nominate a research topic based on a real life sustainability issue of concern for a specific sector of the community or the wider public. The teacher scaffolded the students' research by suggesting resources, ways to manage a very broad or complex topic, or by highlighting any assumptions which needed to be considered. Student peers were also encouraged to offer suggestions. Less confident students could browse this expanding online forum to gain insights and ideas to help them identify a suitable issue, and to guide their research. The students then researched their topics following a framework of general questions to prepare a presentation for educating the affected groups or communities. The common structure of these presentations allowed a direct comparison of the findings, in terms of the scope and reach of each issue, the stakeholders affected, the effectiveness of interventions so far, and the prospects for solutions in the future. Peer assessment of the presentations was a requirement, so that all students needed to attend and share their reflections on the information, expanding their knowledge and exposing their attitudes to a diverse range of real world sustainability issues within a supportive setting.

The tasks were progressively more complex, in terms of Bloom's hierarchy of educational objectives. Initially, students were asked to locate and interpret key sustainability documents outlining relevant international or national policies or positions of government and other influential stakeholders. Moving from information gathering and comprehension, they followed a structured checklist and examples provided online to analyse the sustainability of an unpowered device, and then conducted a sustainability audit of a local café of their own choosing. Now with experience to build their confidence, they analysed the sustainability of a short, local food supply chain based on secondary data they could access from reliable sources on the internet. This sequence of tasks prepared students for an otherwise difficult audit of a real, small-scale commercial processing operation of interest to them. On the dedicated online hub, students could view all choices, with ideas and insights from the teacher and their peers.

They collected information directly from the source through interview and observation, then relied on the higher order skills of synthesis and evaluation, to make an assessment of sustainability along with recommendations for the processor.

Outcomes

Reliance on online technologies for delivery of this course was compatible with teaching to develop capabilities for sustainability in many ways. In the first instance, these technologies had facilitated a dynamic and flexible approach to teaching and learning which was critical for a course dealing with constantly evolving sustainability knowledge. While content and assessment were necessarily prescribed at the outset, it was possible to immediately extend or expand on particular areas of interest to students, and to scaffold learning with new or adapted resources as required. Making adjustments in direct response to students' expressed interests was one way that they could develop a sense of autonomy or control over their learning. By documenting these changes online as they were made, students could keep pace with new information and be confident about the assessment requirements. In the past, it had been the teacher who developed and compiled the learning resources at the outset. Using this model, students were now actively locating, then sharing resources with their own interpretations online. With its reliance on participatory processes and a socially constructed knowledge base as the outcome, a curriculum offering the potential for transformative learning was emerging.

Effective knowledge management is critical for dealing with the messy and unpredictable problems of the future which demand creative and resourceful solutions derived collaboratively and democratically by communities. This was an important theme influencing curriculum design for this project. At the capture stage of the knowledge management cycle, posting all responses online exposed students to much more information and the many different perspectives of their peers. In this online environment, students could directly compare their response to others, providing the initial stimulus for self-reflection. As well as demonstrating the learning due to collaboration, a synthesis of their responses by the teacher guided subsequent student contributions to the next task. The exposure to other ideas and alternative interpretations can create 'cognitive conflict' leading to new understandings (Habron et al., 2012, p. 380), and was critical for promoting transformative learning, an important objective of this curriculum model.

Assessing capabilities for sustainability has been the focus of considerable research. Some capabilities can be readily measured. For instance, it was possible to quantify the number of real world issues to which students were exposed within this course when they attended and reviewed the series of peer presentations. The number of references to social, environmental and economic dimensions of a problem could also be easily counted in each online contribution. Even though the number of 'hits' or messages posted by students has considerable limitations as an indicator of engagement in active learning (Kim, 2013), on-task collaboration between students is somewhat quantifiable when interactions are documented online. Some capabilities are much less amenable to quantitative assessment. In particular, the difficulties in reliably assessing pro-sustainability changes in values and behaviour can confound research in this area. In an effort to deal with this dilemma, a distinction was made between the attributes which would be assessed quantitatively and those which would be assessed qualitatively. In this course, students were assessed for each task using both a numerical scale and descriptive statements as a record of the development of more tangible and less tangible capabilities, respectively.

The success of this project relied on students regularly contributing to the ongoing processes of knowledge collection, exchange, reflection and re-construction. There were some practical hurdles to overcome to motivate students to continue to participate in these processes. For instance, some students enrolled in this course expressed difficulties in connecting with others for collaboration to complete tasks. The need to support students, particularly in the early stages of forming these connections, has been noted elsewhere (Bacelar-Nicolau et al., 2009; Garrison and Kanuka, 2004). In this study, posts were set up to help connect individuals for this work. So, as well as the online task-related interfaces planned by the teacher, here was an example of an interface created to scaffold students' engagement for the necessary collaborations.

Some disciplinary-based assumptions became obvious in class discussions around sustainability. Confronting strongly held convictions and understandings can undermine confidence and compromise self-efficacy, an important capability for sustainability. In line with knowledge management principles, an open and diplomatic discourse in class set the tone and expectations for the online interactions which were to follow. Having the capacity to listen to the concerns and alternative interpretations of around 40 students was critical for engaging everyone in the participatory processes.

Publishing all contributions online is a way of showing respect for their personal perspectives. Providing prompt and personalised feedback also seemed to motivate the students to continue participating, as observed elsewhere (Sridharan et al., 2010). This experience highlighted the 'special challenge' for the teacher in maintaining a 'presence' to facilitate and guide the learning (Garrison and Kanuka, 2004, p. 98). The extent to which teachers can offer this type of feedback to encourage participation in larger classes may be limited, even with the considerable efficiencies afforded by online technologies.

For this model, the blended learning approach was critical for ensuring that virtual interactions were as productive as the real social interactions. Allocating class time for considerable discussion around each upcoming task helped to establish expectations about the quality of contributions to be made at the online interfaces. The use of consistent language, including technical terms, in class and online, was another intentional strategy used to maintain links between the face-to-face discourse and the online task-oriented knowledge exchange and construction.

Conclusions

Orienting university curricula to sustainability requires support from many directions. While a dedicated program and whole-of-university commitment to education for sustainability is the ideal, a single course may be a practical initiative in the interim. However, there are many constraints to adapting an existing course to education for sustainability, including the requirements for academic assessment, students' prior learning experiences and expectations, pedagogical traditions and disciplinary-based thinking. Online technologies offer new opportunities for overcoming these constraints and re-thinking ways to develop curricula for sustainability.

Because sustainability knowledge and other capabilities are the result of social processes, education for sustainability relies on productive human interactions. These interactions rely on communication flow, spaces for interaction and documentation at each stage, conditions which can be readily met by online technologies.

The synergies between knowledge management principles and the ideas of some early educational theorists provided the basis for this curriculum model. Following Bloom's hierarchy and adopting a constructivist approach, a series of tasks involved students in information gathering and comprehension, with more emphasis on application and analysis in later tasks. In between each task, the teacher modelled the synthesis and evaluation of the new knowledge online. The students practiced these higher level cognitive processes in group discussions in each of the following face-to-face sessions, and then autonomously in the final task they completed.

Two important conditions for transformative learning were met by this model. Firstly, it recognised learning in the processes and products of interactions between students and teacher, so distributed the responsibility for learning across all participants. Secondly, the regular cycles of collaborative knowledge construction associated with each online task prompted self-reflection on values and attitudes, a critical step towards transformative learning.

Having access to data describing trends in sustainability-related attitudes, values and behaviours at the start and end of this 12 week course would have allowed validation of the tentative conclusions drawn here. But as much as the findings invite questions, this project has explored another way to understand the possibilities for embedding education for sustainability in university curricula through e-learning. More specifically, it offers ideas for intervening in the ongoing curriculum change processes so that education for sustainability becomes a priority. Further development of this model is contingent on finding reliable indicators of transformative learning in individuals, and particularly for the measurement of social capital, as an outcome of participating in an open, accessible and accepting e-learning environment.

References

Adomssent, M. (2013) Exploring universities' transformative potential for sustainability-bound learning in changing landscapes of knowledge communication. *Journal of Cleaner Production* 49, 11-24.

Adomssent, M., Godemann, J. and Michelsen, G. (2007) Transferability of approaches to sustainable development at universities as a challenge. *International Journal of Sustainability in Higher Education* 8 (4), 385-402

Bacelar-Nicolau, P., Caeiro, S., Martinho, A.P., Azeiteiro, U.M. and Amador, F. (2009) E-learning for the environment. *International Journal of Sustainability in Higher Education* 10 (4), 354-367

Barth, M. (2013) Many roads lead to sustainability: a process-oriented analysis of change in higher education. *International Journal of Sustainability in Higher Education* 14 (2), 160-175.

Barth. M. and Burandt, S. (2013) Adding the "e-" to Learning for Sustainable Development: Challenges and Innovation. *Sustainability* 5, 2609-2622.

Barth, M., Godemann, J., Rieckmann, M. and Stoltenberg, U. (2007) Developing key competencies for sustainable development in higher education. *International Journal of Sustainability in Higher Education* 8 (4), 416-430

Beringer, A. (2007) The Luneburg Sustainable Project in international comparison. *International Journal of Sustainability in Higher Education* 8 (4), 446-461

Cebrian, G., Grace, M. and Humphris, D. (2013) Organisational learning towards sustainability in higher education. *Sustainability Accounting, Management and Policy Journal* 4 (3), 285-306

Clark, B. and Button, C. (2011) Sustainability transdisciplinary education model: interface of arts, science, and community (STEM). *International Journal of Sustainability in Higher Education* 12 (1), 41-54.

Csete, J. and Evans, J. (2013) Strategies for impact: enabling e-learning project initiatives. *Campus-Wide Information Systems* 30 (3), 165-173

de Haan, G. (2006) The BLK '21' programme in Germany: a 'Gestaltungskompetenz'-based model for Education for Sustainable Development. *Environmental Education Research* 12 (1), 19-32

de Kraker, J., Corvers, R., Valkering, P., Hermans, M. and Rikers, J. (2013) Learning for sustainable regional development: towards learning networks 2.0? *Journal of Cleaner Production* 49: 114-122

Dlouha, J., Barton, A., Janouskova, S. and Dlouhy, J. (2013a) Social learning indicators in sustainability-oriented regional learning networks. *Journal of Cleaner Production* 49, 64-73

Dlouha, J., Machackova-Henderson, L. and Dlouhy, J. (2013b) Learning networks with involvement of higher education institutions. *Journal of Cleaner Production* 49, 95-104

Drysdale, J. S., Graham, C. R., Spring, K. J. and Halverson, L. R. (2013) An analysis of research trends in dissertations and theses studying blended learning. *Internet and Higher Education* 17, 90-100.

Ferrer-Balas, D., Lozano, R., Huisingh, D., Buckland, H., Ysern, P. and Zilahy, G. (2010) Going beyond the rhetoric: system-wide changes in universities for sustainable societies. *Journal of Cleaner Production* 18, 607-610

Garrison, D. R. and Kanuka, H. (2004) Blended learning: Uncovering its transformative potential in higher education. *Internet and Higher Education* 7, 95-105

Graham, C. R. and Dziuban, C. (2008) Chapter 23: Blended learning environments. In J. M.
    Spector, M. D. Merrill, J. van Merrienboer and M. P. Driscoll (Eds) *Handbook of re-
    search on educational communications and technology* (3rd Ed) New York: Lawrence
    Erlbaum Associates. 269-276.
Graham, C. R., Woodfield, W. and Harrison, J. B. (2013) A framework for institutional adop-
    tion and implementation of blended learning in higher education. *Internet and Higher
    Education* 18, 4-14
Habron, G., Goralnik, L. and Thorp, L. (2012) Embracing the learning paradigm to foster
    systems thinking. *International Journal of Sustainability in Higher Education* 13 (4),
    378-393
Halverson, L. R., Graham, C. R., Spring, K. J., Drysdale, J. S. and Henrie, C. R. (2014) A
    thematic analysis of the most highly cited scholarship in the first decade of blended learn-
    ing research. *Internet and Higher Education* 20, 20-34
Hesselbarth, C. and Schaltegger, S. (2014) Educating change agents for sustainability – learn-
    ings from the first sustainability management master of business administration. *Journal
    of Cleaner Production* 62, 24-36
Holmberg, J., Lundqvist, U., Svanstrom, M., Arehag, M. (2012) The university and transfor-
    mation towards sustainability. *International Journal of Sustainability in Higher Educa-
    tion* 13 (3), 219-231
Kim, J. (2013) Influence of group size on students' participation in online discussion forums.
    Computers & Education 62, 123-129
Lopez, O. S. (2013) Creating a sustainable university and community through a Common
    Experience. *International Journal of Sustainability in Higher Education* 14 (3), 291-309
Lozano, R. and Young, W. (2013) Assessing sustainability in university curricula: exploring
    the influence of student numbers and course credits. *Journal of Cleaner Production* 49,
    134-141
Macgregor, G. and Turner, J. (2009) Revisiting e-learning effectiveness: proposing a concep-
    tual model. *Interactive Technology and Smart Education* 6 (3), 156-172.
Mahmood, T. and Hafeez, K. (2013) Performance assessment of an e-learning software system
    for sustainability. *International Journal of Quality and Service Sciences* 5 (2), 208-229
Miller, R. (2011) *Vygotsky in perspective.* Cambridge University Press, New York
Miller, T. R., Munoz-Erickson, T. and Redman, C. L. (2011) Transforming knowledge for
    sustainability: towards adaptive academic institutions. *International Journal of Sustain-
    ability in Higher Education* 12 (2), 177-192
Mochizuki, Y. and Fadeeva, Z. (2010) Competencies for sustainable development and sustai-
    nability. Significance and challenges for ESD. *International Journal of Sustainability in
    Higher Education* 11 (4), 391-403.
Moskal, P., Dziuban, C. and Hartman, J. (2013) Blended learning: A dangerous idea? *Internet
    and Higher Education* 18, 15-23
O'Brien, W. and Sarkis, J. (2014) The potential of community-based sustainability projects
    for deep learning initiatives. *Journal of Cleaner Production* 62, 48-61
Owston, R. (2013) Blended learning policy and implementation: Introduction to the special
    issue. *Internet and Higher Education* 18, 1-3
Owston, R., York, D. and Murtha, S. (2013) Student perceptions and achievement in a uni-
    versity blended learning strategic initiative. *Internet and Higher Education* 18: 38-46
Pappas, E., Pierrakos, O. and Nagel, R. (2013) Using Bloom's Taxonomy to teach sustaina-
    bility in multiple contexts. *Journal of Cleaner Production* 48, 54-64

Peer, V. and Stoeglehner, G. (2013) Universities as change agents for sustainability – framing the role of knowledge transfer and generation in regional development processes. *Journal of Cleaner Production* 44, 85-95

Posch, A. and Steiner, G. (2006) Integrating research and teaching on innovation for sustainable development. *International Journal of Sustainability in Higher Education* 7 (3), 276-292

Rieckmann, M. (2012) Future-oriented higher education: Which key competencies should be fostered through university teaching and learning? *Futures* 44 (2), 127-135

Ross, V. (2012) From transformative outcome based education to blended learning. *Futures* 44, 148-157

Salmons, J. (2009) Chapter 19: E-social constructivism and collaborative e-learning In J. Salmons & L. Wilson (Eds). *Handbook of Research on Electronic Collaboration and Organizational Synergy*. Hershey, PA: IGI Global. Pp 280-294. doi:10.4018/978-1-60566-106-3

Savageau, A. (2013) Let's get personal: making sustainability tangible to students. *International Journal of Sustainability in Higher Education* 14 (1), 15-24

Seraj, M. (2012) We create, we connect, we respect, therefore we are: Intellectual, social and cultural value in online communities. *Journal of Interactive Marketing* 26, 209-222.

Sheate, W. R. and Partidario, M. R. (2010) Strategic approaches and assessment techniques – Potential for knowledge brokerage towards sustainability. *Environmental Impact Assessment Review* 30, 278-288

Shephard, K. (2008) Higher education for sustainability: seeking affective learning outcomes. *International Journal of Sustainability in Higher Education* 9 (1), 87-98

Sridharan, B., Deng, H. and Corbitt, B. (2010) Critical success factors in e-learning ecosystems: a qualitative study. *Journal of Systems and Information Technology* 12 (4), 263-288

Stubbs, W. and Schapper, J. (2011) Two approaches to curriculum development for educating for sustainability and CSR. *International Journal of Sustainability in Higher Education* 12 (3), 259-268

Tello, G., Swanson, D., Floyd, L. and Caldwell, C. (2013) Transformative learning: A new model for business ethics education. *Journal of Multidisciplinary Research* 5 (1), 105-120

Wals, A. (2014) Sustainability in higher education in the context of the UN DESD: a review of learning and institutionalization processes *Journal of Cleaner Production* 62, 8-15

Wiek, A., Withycombe, L. and Redman, C. L. (2011) Key competencies in sustainability: a reference framework for academic program development. *Sustainability Science* 6, 203-218

Wiek, A., Bernstein, M. J., Laubichler, M., Caniglia, G., Minteer, B. and Lang, D. J. (2013) A global classroom for international sustainability education. *Creative Education* 4 (4A), 19-28

Wood, D., Bruner, J. and Ross, G. (2006) Chapter 16: The role of tutoring in problem solving. In J. Bruner (Ed) *In search of Pedagogy. Volume 1: The selected works of Jerome Bruner*, 1957-1978. Routledge – Taylor and Francis Group, New York

Yu, F. (2009) Scaffolding student-generated questions: Design and development of a customizable online learning system. *Computers in Human Behaviour* 25, 1129-1138.

E-learning for sustainable development: linking virtual mobility and transboundary competence development

Joop de Kraker[1,2], Ron Cörvers[2] and Angelique Lansu[1]

Abstract

More than twenty years ago, Agenda 21 called for the reorientation of education towards sustainable development. In this context, there is a growing attention for the concept of 'competences for sustainable development'. These competences describe the desired output of education for sustainable development (ESD) and are thus important determinants of how education should be reoriented and restructured. In this chapter, we review how these competences have been defined for higher education and argue that a key competence to be developed is 'transboundary competence'. Transboundary competence is the ability for productive interaction across the boundaries between different perspectives, for example boundaries between scientific disciplines or between scientific and other types of knowing. We then apply the principles of competence-based learning to derive the design principles for effective learning environments to foster transboundary competence. Important design principles are cross-boundary contexts and collaborative learning in heterogeneous groups. We argue that these requirements can be met by employing virtual mobility. Virtual mobility, i.e., using web-based learning platforms to bring geographically distributed students and teachers together, allows the creation of culturally, disciplinary and experientially diverse learning teams. We illustrate this approach to fostering transboundary competence through virtual mobility with two successful examples. The examples show that this e-learning approach to ESD is relevant for both open distance teaching universities and regular universities.

Introduction

Ever since Agenda 21 called for the reorientation of education towards sustainable development (UNCED, 1992), numerous activities have been undertaken worldwide to achieve this goal, most notably during the UN Decade of Education for Sustainable Development from 2005 till 2014 (UNESCO, 2005). In 'The Future We Want', the outcome document of the Rio+20 conference, the UN member states affirmed their commitment to continue promotion of Education for Sustainable Development (ESD) beyond the Decade (UN, 2012). The call for ESD has been echoed in many policy documents, for example in the EU Sustainable Development Strategy, which emphasizes that education should provide all citizens with the key competences needed to achieve sustainable development (European Council, 2006). Competences, which the EU defines as 'combinations of knowledge, skills, and attitudes appropriate to the context' (European Commission, 2006), describe the desired 'output' of

---

1   School of Science, Open Universiteit, Heerlen, The Netherlands.
2   ICIS, Maastricht University, Maastricht, The Netherlands.

ESD and are thus important determinants of how education should be reoriented and restructured.

Despite this important role, high-level policy documents on ESD do not specify what important competences for sustainable development are (Mochizuki and Fadeeva, 2010). Apparently this task is left to the educational sectors to perform. In the higher education sector, however, remarkably little attention has been paid to the issue of competences for sustainable development: e.g., only 2% of the papers published in the International Journal of Sustainability in Higher Education between 2000 and 2010, focused on competences, whereas 25% addressed the issue of environmental management of university campuses (Wals and Blewitt, 2010). Nevertheless, interest in competences for sustainable development is rapidly growing, as is illustrated by the steep increase in the number of publications per year on this topic: a search in Google Scholar on publications with these keywords in the title between 2000 and 2013, yielded a harvest of, on average, 3 publications per year during the first five years of this period, to over 25 per year during the past five years.

The common approach to defining competences for sustainable development is to derive these from the typical characteristics of a sustainability problem: complex, having multiple dimensions and spanning multiple scales in time and space, surrounded by uncertainties, and involving a broad range of actors representing a diversity of perspectives on the issue. The latter characteristic makes clear that the ability to deal with diverse or even diverging perspectives must be an important competence for sustainable development. We have labeled this ability 'transboundary competence', the competence to interact across the boundaries between different perspectives in a productive way (De Kraker et al., 2007).

In this chapter, we first review the 'state-of-the-art' concerning definition of competences for sustainable development in higher education and we elaborate in particular on transboundary competence. We then consider the development of transboundary competence in the context of e-learning. By applying the principles of competence-based learning, we derive the requirements for e-learning environments to foster transboundary competence, and argue that these requirements can be met by linking transboundary competence development to virtual mobility, i.e., using web-based learning platforms to bring geographically distributed students and teachers together without the need to travel. We illustrate this approach with two successful examples, and end the chapter with a discussion and outlook on further developments.

Competences for sustainable development

Specifying key competences

The debate about competences for sustainable development that has been going on over the past five to ten years, is characterized by a great diversity in terminological meaning, orientation and approach.

Almost every term of the concept of 'key competences for sustainable development' is contested, up to the level of correct spelling (competences versus competencies). In the Anglo-Saxon world, 'competences' are associated with concrete skills that may be part of national vocational standards, whereas in continental Europe and EU policy, competence is a more holistic concept (Van der Klink and Boon, 2003). In the latter case, the term 'key' is used to indicate that a given competence is cross-cutting and important for every member of society (De Haan, 2006; European Commission, 2006), but other contributors to the debate use the term 'key' (or 'core') to indicate almost the opposite, namely that a competence is critical to professional success and distinctive for a particular domain (Wiek et al., 2011). Usage of seemingly interchangeable terms like 'Education for Sustainable Development', 'Learning for Sustainable Development', 'Education for Sustainability', 'Sustainability Education' and 'Sustainable Education', is associated with large differences in meaning (Thomas, 2009). This comes as no surprise when one considers the value-laden nature of the concepts involved. Given the diversity in the terminology, every contribution to the debate should provide clear definitions in order to be useful.

A similar diversity can be observed in the orientation of the competences, either towards the student as a person, with a focus on values and behaviour, as a citizen, with a focus on participation in societal decision making, or as a professional, with a focus on future contributions to sustainable development based on specific expertise. Of course combinations are possible, but for higher education, the emphasis is usually on the competences students will need as a professional. In that case, competences may be defined for graduates of all disciplines (Barth, 2007; De Kraker et al., 2007; Roorda, 2010; Rieckmann, 2012; Thomas et al., 2013), for engineers (Segalas et al., 2009; Hanning et al., 2012), or for various specialists in sustainable development, such as teachers (UNECE, 2012), coordinators (Willard et al., 2010), entrepreneurs (Lans et al., 2013), mediators (Johansson and Laessoe, 2008), and sustainability researchers and problem-solvers (Wiek et al., 2011).

To specify important competences for sustainable development, the common approach is to derive these from the typical characteristics of a sustainability problem: complexity, uncertainty and a diversity of perspectives (e.g., De Haan, 2010; Wiek et al., 2011). Another common approach is to first compile long-lists of potentially relevant competences and then ask experts, professionals, employers, or alumni to select a short-list of competences deemed most important (e.g., Willard et al., 2010; Hanning et al., 2012; Rieckmann, 2012; Thomas et al., 2013). In these approaches, sometimes a clear distinction is made between competences for sustainable development on the one hand, and generic academic and domain-specific competences on the other hand, sometimes this is not clearly separated.

Our position amidst all this diversity in terminology, orientation and approach, is as follows. We define competence as 'an integrative whole of knowledge, skills and attitudes, applied to perform complex tasks in authentic work environments' (De Kraker et al., 2007). With the term 'key competences for sustainable development' we refer to competences deemed critical for academic professionals to contribute effectively to changes towards a more sustainable society, considering the economic, environmental as well as the socio-cultural dimension, current and future generations, both locally and globally. These key competences are oriented towards students as future professionals in all academic disciplines ('academic professionals'). In our view, specification of key competences should primarily be based on an analysis of the role that academic professionals can and do play in change processes towards a more sustainable society.

The need for transboundary competence

In the previously cited body of literature on competences for sustainable development in higher education, several tendencies are apparent that we consider problematic. These concerns:

- an overestimation of the possibility for students to acquire a vast range of complex competences in addition to generic academic and domain-specific competences;
- an underestimation of the complexity of societal change processes in combination with an overestimation of the role of individual 'change agents';

- an overestimation of the rationality of public policy and decision making, in combination with an overestimation of the role of scientific knowledge in such processes; and finally,
- an overestimation of the possibility for academic professionals to develop a single over-arching, holistic perspective of a complex societal problem, in combination with an underestimation of (irreducible) plurality in perspectives held by the many actors involved in such problems.

Other scholars, notably Vare and Scott (2007), have already pointed at the lack of attention for the fundamental uncertainties in sustainable development that is apparent in a large part of the literature. This issue is known as the controversy between the 'instrumentalist' and the 'emancipatory' approach to ESD (Wals and Jickling, 2002), but as these labels are rather suggestive, we prefer to call them neutrally ESD1 and ESD2, following Vare and Scott (2007). As the latter argue, the essential difference between ESD1 and ESD2 is that ESD1 assumes that the right diagnosis and effective solutions of unsustainability are known, whereas ESD2 stresses that these are surrounded by fundamental uncertainties given the complexity of sustainability issues and the unpredictability of future developments. In ESD2, therefore, independent and critical thinking about diagnoses and proposed solutions of sustainability issues is considered the most important competence for learners to acquire (Jickling, 1992; Wals and Jickling, 2002; Vare and Scott, 2007). We agree with the criticism on the lack of attention for uncertainty in ESD1, which is also reflected in the key competences specified in the ESD1 literature. Development of the competence of independent, critical thinking, however, we consider an academic conditio-sine-qua-non, and not an additional key competence specifically required to make constructive contributions to the solution of sustainability problems.

In our view, both ESD1 and ESD2 do not pay sufficient attention to the fact that in dealing with complex, 'wicked' sustainability issues, academic professionals will always have to collaborate with many other actors, both academic and non-academic, bringing with them a broad diversity in knowledge and perspectives on the issue (Cash et al., 2003; Gallopin et al., 2011). This diversity follows from the ambition underlying the concept of sustainable development to link the ecological, economical and socio-cultural aspects of development, and to connect development here and now with development elsewhere and in the future. This multi-dimensional approach to development is so comprehensive that it is (almost) impossible to catch it in a single, balanced perspective. Furthermore, due to the complex nature of sustainability problems, the large spatial scales on which they occur and their long-term effects, there is considerable uncertainty and (consequently) dissensus in the scientific knowledge about these problems and their solutions. This leaves room for multiple interpretations of the risks involved and how they should best be managed. Finally, sustainable development is in essence a normative concept, and although there is a broad agreement on its core values of inter- and intra-generational justice, the interpretation and the relative weight that should be given to the various interests when weighing the pros and cons will differ according to one's life philosophy, social position, discipline, etc.

The diversity of perspectives can be valued positively for several reasons. The thrust of sustainable development is to prevent as much as possible shifting of the burden of improvements in one domain or for one group to other domains or groups. Taking a diversity of perspectives into account will thus provide a sharper eye to detect such undesirable shifts, and may result in more balanced decisions. Multiple perspectives could also enable a richer definition of complex sustainability problems and produce a wider array of potential solutions, which, in the face of uncertainty, enhances the probability to find adequate solutions (Janssen and Osnas, 2005).

However, the diversity of perspectives also entails a risk of conflict, political paralysis and a lack of societal support at a time when joint, large-scale measures may be urgently needed (Keulartz, 2005). Such situations are likely to occur, because, despite intentions to prevent unjust shifts of costs, some groups are bound to win (or loose) more than others. Diversity in perspectives thus creates the need for negotiation, dialogue and learning between the actors involved in a sustainability issue (Leeuwis, 2002), to arrive at richer, more complete definitions of sustainability problems, a wider array of potential solutions, and more balanced, broadly supported strategies. We therefore see the ability to interact across the boundaries of diverse perspectives in a productive way, as a key competence for sustainable development, to be acquired by students of all disciplines in higher education. We refer to this ability as 'transboundary competence'.

The concept of transboundary competence

In the context of higher education, we define competences for sustainable development as those combinations of knowledge, skills and attitudes that enable graduates of all disciplines as academic professionals to effectively contribute to transition processes towards a (more) sustainable society, on the basis of their domain-specific expertise and more general academic competences. As outlined above, transboundary competence, the ability to deal with a diversity of perspectives in a productive way, is crucial in such societal transition processes. Most graduates will, when contributing to sustainable development, work in teams or networks, on the basis of their specialised expertise, and in the search for sustainable solutions, their scientific contributions will need to be combined with the knowledge, values and interests of other actors.

When multiple actors engage in complex problem solving, there are many boundaries to be crossed (Cash et al., 2002). Major boundaries are those arising from differences in terminology, values or interests (Carlile, 2004). In the practice of sustainable development, various boundaries often co-occur. For example, Valkering et al. (2013), observed in a major transnational initiative on sustainable urban development a mix of boundaries arising from differences in sectoral, societal or territorial backgrounds of actors. These boundaries more or less correspond with those to be crossed in interdisciplinary, transdisciplinary and internationally or interculturally collaboration in problem solving (De Kraker et al., 2007; Roorda 2010). The competence to cross such boundaries, requires certain knowledge, skills and attitudes (Keulartz, 2005; De Kraker et al., 2007; Wals, 2010). In terms of knowledge, awareness of the diversity of perspectives and understanding of the origin of this diversity is needed. Skills include the ability to reflect on one's own perspective and to articulate it, to take (temporarily) someone else's perspective and to negotiate a shared perspective or frame of reference. The required attitudes towards other perspectives include acceptance of their legitimacy, willingness to engage, and belief in the added value of looking at a problem from different perspectives.

Interaction across the boundaries of perspectives normally concerns a two-way communication process, during which a shared frame of reference is established. In the cognitive sciences, this is called 'common ground' and the process of establishing it 'grounding'. Common ground is the sum of knowledge, beliefs and assumptions shared by different perspectives, and, by acting as an interface, forms the basis for a productive dialogue between these perspectives (Bromme, 2000). Common ground makes it possible to understand what the other means, which may be followed by, but is not identical to agreeing with the other, as one may 'agree to disagree'. Beers et al. (2006) describe the process of grounding as a negotiation process, an iterative communication process during which the parties involved articulate their perspectives and give and receive feedback to arrive at mutual understanding ('negotiation of meaning') and – perhaps – at mutual agreement ('negotiation of position'). According to Bromme (2000), the most important basis for developing common ground between different perspectives, is common experience and common activity, which can be reinforced by reflection on the process. Grounding involves the establishment of a common language which may include new, shared concepts. Establishment of common ground usually does not result in entirely 'dissolving' of the differences between perspectives, partly because precisely these differences can make the communication between perspectives worthwhile and productive (Bromme, 2000). Common ground then forms the basis on which solutions can be jointly 'constructed' that integrate or respect the diversity of knowledge, viewpoints, values and interests of the range of actors involved in a complex problem.

Competence development: linking transboundary competence and virtual mobility

If transboundary competence development is important for ESD in higher education, then what are appropriate didactic frameworks and effective learning environments? In this section we will first discuss the concept of competence-based learning and then apply this concept to identify the characteristics of learning environments in which transboundary competence can be developed.

Principles of competence-based learning

The concept of competence-based learning has been developed over the past two decades and has gained ground as a dominant educational approach in Western Europe. Its development was a response to the requests of the labour market and the wish of educators to make formal education more meaningful and relevant. The idea is that learning should focus on integrative competences required in professional life, and not on the acquisition of isolated skills and pieces of knowledge. Life-long learning skills are seen as an essential part of competences, as both the 'state-of-the-art' and professional requirements are rapidly changing (European Commission, 2006).

Constructivism is usually the dominant paradigm in competence-based learning, and learning environments that stimulate active, contextual construction of knowledge and understanding and active acquisition of competences are favoured (Könings et al., 2005). The best way to acquire these competences is in a learning environment that combines actual practice ('learning by doing'), and explicit reflection on what and how to learn from that practice ('learning by reflection').'Learning-by-doing' involves that the learning environment is realistic or authentic in terms of the problems the students have to solve, the tasks they have to perform, and the context of these tasks. 'Learning-by-reflection' involves that students explicitly reflect on their learning goals, activities, results and ways to improve. In our view, the concept of competence-based learning offers an appropriate didactic framework for higher education for sustainable development, in which the acquisition of transboundary competence is seen as central. The constructivist paradigm, which considers knowledge as a (social) construct of reality and not as the absolute objective truth about reality, corresponds with a pluralistic approach to the multiple perspectives in society on sustainable development, whereas the constructivist approach to learning as the (joint) production of context-embedded knowledge matches the need to find context-specific sustainable solutions shared and supported by many. Moreover, acquiring transboundary competence through a combination of 'learning by doing' and 'learning by reflection', corresponds with Bromme's (2000) observations that the most important basis for developing common ground between different perspectives, is common experience and common activity, which can be reinforced by reflection on the process.

Learning environments fostering transboundary competence development

In the philosophy of competence-based learning, a learning environment for sustainable development should foster transboundary competence through 'learning-by-doing' combined with 'learning-by-reflection' (table 1). In the ideal learning environment, the problems or cases should be ill-structured, and concern sustainability issues that cover multiple domains and span multiple scales of space and time. The tasks to be performed must be open-ended and requiring active integration of diverse knowledge, e.g., from a variety of disciplinary, geographical, or cultural backgrounds.

The tasks should include the need to consider different interest- and value-based perspectives on the problem. A realistic context could be created by introducing teamwork in heterogeneous groups in the learning environment and/or by creating an open learning environment. This would necessitate the students to communicate or even collaborate with real actors with a different disciplinary, national or cultural background or societal position. Essential for effective competence-based learning is explicit reflection on the activities, how they contribute to the development of transboundary competence, how this relates to personal goals, and how performance and learning could be improved. This would involve writing and discussion of reflection reports. In this way, the students also learn better to reflect, which is an essential skill for crossing the boundaries of one's own perspective. For the development of transboundary competence it is important not only to reflect on the quality of content-related products (such as problem definitions, analyses, solutions), but also on the processes or strategies by which these products came about. In particular, processes like negotiation of meaning and position in heterogeneous groups should be a topic of explicit reflection. Individual reflection must be complemented by the perspectives of others through group discussions of these topics and organised feedback, preferably also from outside the university.

Table 1: Characteristics of a competence-based learning environment applied to transboundary competence development

General feature	Application to transboundary competence development
realistic problems or cases	• ill-structured problem description • multiple scale, multiple domain issues
realistic tasks or roles	• open-ended • active integration of different aspects of problem and knowledge from different backgrounds • dealing with different interest- and value-based perspectives on the problem
realistic context	• heterogeneous student groups (multidisciplinary, multi-cultural, international) • open learning environment, interaction with experts or stakeholders from outside university
explicit reflection on task performance and learning	• reflection on quality of products in relation to quality of processes and learning strategies • reflection on processes in heterogeneous groups (negotiation, learning) • individual reflection complemented by group discussions and organized feedback (preferably also from outside the university)

After: De Kraker et al. (2007).

Virtual mobility

In a traditional university setting, a learning environment with group work on projects in realistic, cross-boundary contexts is difficult to realise. It would require a high level of international
student mobility to bring students from different disciplinary, national and cultural backgrounds
frequently together at the same time and in the same place. Virtual mobility, using web-based
learning environments, provides an innovative solution to this problem. With modern ICT,
geographically distributed students and teachers can be brought together without the need to
travel, allowing the creation of culturally, disciplinary and experientially diverse learning teams
at low cost. For transboundary competence development, an added advantage is that virtual
learning environments can provide better opportunities to structure the group discussions and
reflection processes, both individually and collectively, because most communication will be in
writing (Barth, 2007). Certain tools of virtual learning environments, such as a Wiki space,
foster a constructivist approach to knowledge and support dealing with complex concepts from
multiple angles (Burandt and Barth 2010), both of which contribute to the development of transboundary competence. Thus, although it might seem somewhat counter-intuitive, virtual mobility combined with web-based learning environments offers an effective framework for development of transboundary competence. In the next section, we illustrate this approach with two
successful examples: the European Virtual Seminar on Sustainable Development (EVS) and
the 'Lived Experience of Climate Change – e-learning and virtual mobility' (LECH-e) pilot.

Two examples: EVS and LECH-e

In presenting EVS and LECH-e as examples of how purpose-designed virtual learning environments can be combined with virtual mobility to foster development of transboundary
competence, the focus will be on the features of the learning environment (following table 1)
and on the structuring of the group learning process. Our account of the development of
transboundary competence in both examples is based on team reflection reports and formal
assessments of the team collaboration process and team reports in EVS, and on student reflection reports and a pilot evaluation questionnaire in LECH-e. More extensive descriptions
can be found for EVS in Cörvers and De Kraker (2009) and De Kraker and Cörvers (2014),
and for LECH-e in Wilson et al. (2011), and Perez-Salgado et al. (2012, 2014).

EVS

The European Virtual Seminar on Sustainable Development (EVS, *www.ou.nl/evs*) is a Mas-
ter-level university course with a study load of 120 hours. In EVS, students work together on
case studies in sustainable development in international, multi-disciplinary virtual teams.
Each team of 4-8 students is coached by a tutor, focusing on the group process, and an expert
on the case study topic, focusing on the group products. Students and staff communicate
mainly asynchronously, using a tailor-made web-based platform, supplemented with real-
time team sessions using Skype. EVS has been running as an annual course since 2001, co-
organized by staff from 12 universities, with an enrolment rate of about 60 students per year.
Given that the EVS students come from 10 different countries from across Europe, and from
a range of degree programs, the geographic, cultural and disciplinary diversity of the student
teams is usually very high. This diversity is further amplified by the participation of students
from three open universities, who are usually substantially older than the students from regular universities and combine their studies with a professional life.

Development of transboundary competence is a major goal of EVS and both the learning environment and the learning process have been designed and structured to promote this development following the principles of competence-based learning (tables 2 and 3). The overall theme of EVS is sustainable development in Europe. An EVS case study therefore focuses on issues in sustainable development (e.g., energy, waste, climate change) in one or more European countries. Such a case study is not an elaborate description of a specific case with a predefined problem, but a brief, open-ended problem description requiring further study. The ultimate objective of the study is to come up with recommendations on how to promote sustainable development for the topic of the case study. The challenge is to make use, as much as possible, of the diversity in the team in terms of different perspectives on the problem and the variety of knowledge and expertise of its members. This diversity is relevant, because not only the various dimensions of the case study topics have to be addressed, but also a comparative approach is encouraged, usually by studying how the issue is dealt with in the home countries of the team members.

Table 2:   Features of EVS as a learning environment for transboundary competence development

General feature	EVS
realistic problems or cases	• brief problem or case descriptions • sustainability issues • European perspective
realistic tasks or roles	• open-ended case study research • active integration of different dimensions of problem and knowledge from different disciplinary, national, cultural and professional backgrounds • dealing with different member state-perspectives on the problem
realistic context	• heterogeneous student teams (multidisciplinary, multi-cultural, international) • semi-open learning environment: interaction with experts from other universities and disciplines
explicit reflection on task performance and learning	• reflection on individual learning process • reflection on group processes in heterogeneous team (negotiation, learning) • individual reflections complemented by group reflection

As EVS is spread out over a period of five months, there is ample time for the teams to develop as they go through the various stages of the course, and observations and reflection reports make it invariably clear that the team members grow in transboundary competence. Bit by bit, through often intense discussions, they negotiate common ground upon which they build joint knowledge, most tangibly in the form of written group products (table 3). During the first group activities the emphasis is on grounding discussions and during the later activities more on knowledge integration.

The level of transboundary competence the teams have reached by the end of EVS differs considerably, depending most strongly on the willingness of the students to invest in the group process and the ability of the tutor to motivate them to do this.

Table 3:   Structure of the group learning process in EVS.

Activities	Products
1. Introducing yourself	• personal profile page • introductory message
2. Discussing the basic concepts	• shared definition of sustainable development, from a European perspective, applied to the topic of the case study
3. Inventory of team diversity	• overview of expectations, interests, knowledge and skills relevant to the topic of the case study
4. Discussing the research objectives	• shared research objective(s) for the case study
5. Individual reflection	• brief, forward-looking report on individual expectations, learning goals and first experiences
6. Developing the research proposal	• commonly agreed plan describing what, how, and when to research the case and the roles and tasks of each team member
7. Conducting case study research	• individual contributions to the group research report
8. Writing the group research report	• coherent, integrated final report with a shared set of conclusions and recommendations
9. Individual & group reflection	• retrospective individual report on EVS experiences, fulfillment of expectations, and learning achievements • evaluative group report on group process and achievements, and possible improvements of EVS

LECH-e

In the 'Lived Experience of Climate Change – e-learning and virtual mobility' project (LECH-e, *www.leche.open.ac.uk*), learning materials and a virtual learning environment were developed for a complementary Masters track on climate change in the context of sustainable development. This EU-funded project was conducted from 2009 to 2012 by a consortium consisting of five open distance teaching universities and three regular universities. Delivery of the Masters track with a study load of 360 hours was piloted in 2011, involving about 30 students from six European universities. The diversity of these students in terms of backgrounds was high, and included six different degree programs, more than 10 different nationalities and over 30 different disciplines and areas of expertise. As the pilot was very successful and positively evaluated, the LECH-e track has since 2011 been offered by the German open university (FernUniversität), with at least three other universities intending to join.

The learning goals of the LECH-e track focus on awareness of the diversity of perspectives on the issue of climate change and appreciation of the value of contextual knowledge gained through everyday experience, alongside knowledge on climate change from both the natural and social science disciplines, to inform policy making and action. Given these goals, transboundary competence was considered an important ability for the students to develop, and the learning environment and learning process were designed to stimulate this development (tables 4 and 5). In the LECH-e pilot, the students were expected to spend more than 50% of their study time on workbook activities, which involved applying the concepts discussed in the textbook to case studies, developing integrated views on the basis of a variety of (disciplinary) sources and discussing these with fellow students in web-based forums, moderated by staff.

Students were made aware of the diverse perspectives on climate change that originate from the diversity of geographical, cultural, social and economic conditions under which people live, by providing them with or making them actively look for accounts of lived experience of climate change from a variety of contexts. There were three formal group activities in which students had to interact across the boundaries of different perspectives, i.e., practice transboundary skills (table 5). In the final assessment paper, the students were required to reflect critically on the value of transboundary competence with respect to climate change, drawing on the discussions they had with fellow students during the group activities (accounting for 20% of the final mark).

The time available for the three formal group activities in the LECH-e pilot was about one month, which, according to the students, was too short for in-depth communication and development of transboundary competence in a more profound way. In their reflections on transboundary competence, the students demonstrated their awareness of the diversity of perspectives on climate change and confirmed the importance of dealing with this diversity in a productive way. The reflections generally contained a rich and deep analysis of the processes in the group activities, and an insightful assessment of the factors promoting and blocking transboundary engagement.

Table 4: Features of LECH-e as a learning environment for transboundary competence development

General feature	LECH-e
realistic problems or cases	• medium-structured problem descriptions • climate change issues
realistic tasks or roles	• perspective-based cost-benefit analysis; preparation of joint policy brief • active integration of knowledge from different perspectives • dealing with different interest- and value-based perspectives on the problem in multi-actor negotiations (role play)
realistic context	• heterogeneous student groups (multidisciplinary, multi-cultural, international) • role play including realistic diversity of actor perspectives
explicit reflection on task performance and learning	• reflection on transboundary competence, drawing on the group processes and outcomes

The authenticity of the various perspectives in the three group activities varied strongly: in the carbon footprint activity, perspectives were most authentic (i.e., genuine, real), namely the students' own, whereas in the flood defence activity the perspectives were least authentic, because this was a role play in which the students had to take predefined perpectives. In the policy brief activity, the students could indicate their preference for a certain perspective to take, so in this case the authenticity of the perspectives was intermediate. The students felt that practicing transboundary competence would be more effective when the perspectives to be taken would be more authentic (i.e., closer to their own, personal perspective), which is in line with the concept of competence-based learning, and an important guideline for designing learning environments for transboundary competence development.

Table 5: Formal group work activities in LECH-e

Activities	Outcomes
1. Calculation and comparison of personal carbon footprints	• Exchange of carbon footprints served as an ice-breaker for students who had never met. • Rich discussion and reflection on how different lifestyles and contexts may impact on climate change.
2. Cost-benefit analysis of a river flood defence from different perspectives (environmentalists, inhabitants, project developer), with the aim to negotiate a joint advice to the local authority	• No common agreement was reached between the three groups. • Students experienced the difficulties in negotiating a joint, integrative solution, as the perspectives were to a large extent oppositional.
3. Preparation of a collective policy brief on the relevance of lived experience for EU climate change policy, incorporating a variety of perspectives (focusing on loss of amenity and landscape, livelihood impacts in EU and developing countries, and communication of climate science)	• A high quality, integrated document was produced. • Students experienced conditions favouring transboundary collaboration, as the four perspectives were diverse but not necessarily divergent and shared a common ambition to show the relevance of lived experience.

Discussion

In this chapter, we have highlighted the need for transboundary competence development and focused on the question how the development of this competence can be promoted in higher education, in particular in relation to virtual mobility. This does not mean, however, that in the ESD debate we are unique in indicating the need for transboundary competence development, nor that we are the only ones that have specified transboundary competence. The importance of the ability to communicate with a broad range of stakeholders and across disciplines and to understand and respect a variety of perspectives, also came up in other studies on competences for sustainable development (Barth et al., 2007; Newport, 2008; Segalas et al., 2009; Willard et al., 2010; Hanning et al., 2012; Rieckmann et al., 2012; Thomas et al., 2013). Most of the proposed comprehensive frameworks of competences for sustainable development, include (sub) competences resembling or overlapping with transboundary competence (e.g., De Haan 2006, 2010; Roorda, 2010; UNECE, 2011; Wiek et al., 2011). These frameworks also include other key competences relevant to complex problem solving, such as the ability to take a systems approach, to take a future-oriented, long-term perspective and to deal with risk and uncertainty. Our position is that we fully acknowledge the importance of these other competences, but find these more appropriate for ´sustainable development specialists' than for all academic professionals. We argue that for academic professionals of any kind, the most important competence to enable them to contribute to sustainable development in collaboration with other actors, is the ability to interact across the boundaries of diverse perspectives in a productive way. We consider it not realistic for students of all academic disciplines to acquire a whole range of competences for sustainable development, in addition to generic academic competences and domain-specific competences, hence our plea to focus on transboundary competence.

Ideally, development of transboundary competence through a combination of learning by doing and reflection, should be complemented with the acquisition of insight into the roles that academic professionals and scientific knowledge can and do play in addressing complex societal challenges (for an example, see Tuinstra et al., 2013). It is our impression that students of most academic programs tend to overestimate the role of individual 'change agents', the use of scientific knowledge, the possiblity of a single overarching perspective and the rationality of decision making in complex societal change processes. A more realistic view of these issues would help students to better understand the need for transboundary competence, but also make them more aware of the limitations to finding sustainable solutions that integrate or respect the diversity of actor perspectives. For example, in cases when there are deep conflicts between perspectives in terms of values or interests, general transboundary competence will probably not suffice to overcome the boundaries between these perspectives and assistance by specialized mediators could be helpful. Examples of pragmatic strategies mediators could employ to overcome boundaries between perspectives include 'reframing', i.e., redefining the problem at a level where there are common interests, 'gradualisation', i.e., redefining the problem in less dualistic normative terms, the use of 'boundary objects', i.e., shared concepts or models that are flexible enough to adapt to the needs of the various perspectives, and making the boundaries ´selectively porous' for some issues while keeping them solid for other, more contentious issues (Cash et al., 2002; Keulartz, 2005).

In the context of e-learning for sustainable development, we have emphasized the synergy between transboundary competence development and virtual mobility.

Virtual mobility allows the creation of learning environments fostering the development of transboundary competence through collaborative learning in culturally, disciplinary and experientially diverse student groups. Through this role in the development of transboundary competence, virtual student mobility gains an independent and distinctive quality as compared to physical student mobility. It is our experience that such an 'educational necessity' is a critical factor in the establishment of viable forms of virtual mobility (Cörvers and De Kraker, 2009). However, it is also our experience, for example in the LECH-e project, that highly internationalised universities with a very diverse student population, such as Wageningen University in the Netherlands, prefer and are capable to create such heterogeneous learning environments in a face-to-face setting, without having to resort to virtual mobility (cf. Wals, 2010).

For the many universities that do not match these criteria, virtual mobility can be an effective solution, as we demonstrated in the EVS and LECH-e examples. Nevertheless, virtual mobility requires considerable interuniversity coordination and will not be employed in more than one or a few courses in an academic program. To promote the development of transboundary competence also in other courses, the approach proposed by De Vries (2013) might be an effective didactic strategy. In his textbook 'Sustainability Science', the author introduces four archetypical perspectives or world views, ways to see the worlds (cf. De Vries and Petersen, 2009), and invites students to take and reflect on these different perspectives by providing a range of perspective-based statements on each of the sustainability issues discussed in the book.

Outlook

For the future, we foresee growing attention for transboundary competence as a key competence for sustainable development, as well as for employing virtual mobility to create effective learning environments for the development of transboundary competence. The case for transboundary competence development is increasingly taken up by actors outside the ESD community. An important example is a recent report from the European Science Foundation focusing on the contribution of science to the societal challenge of global environmental change (ESF, 2012). The report emphasizes 'skills and abilities of scientists' similar to transboundary competence, such as 'humility and openness towards other disciplines, worldviews and other sources of knowledge, both formal and informal', 'being able to communicate in real (multiple-way) dialogues', and 'willingness to acknowledge that the partial knowledge the researcher brings to the dialogue table will be transformed in the discussion process (giving latitude to other people)'. Also within the ESD community, transboundary competence is more and more considered a key competence. For example, the new Dutch national programme on Learning for Sustainable Development focuses on social innovation and the formation of regional sustainability networks. In terms of competences, the emphasis is on abilities to collaborate and to connect in a productive way (EZ, 2013).

In this chapter, we gave two examples of employing virtual mobility of European students in the development of transboundary competence. There is no reason, however, to limit virtual mobility to within Europe. Barth and Rieckmann (2009) have demonstrated that a virtual seminar involving European as well as Latin American students was effective in improving transboundary competence, and recently Wiek et al. (2013) reported on a 'Global Classroom' virtual mobility initiative to equip German and North-American students with transboundary competence.

Further development of didactic strategies and learning environments for transboundary competence could benefit from the insights of educational scientists and technologists into competence-based learning. According to these insights, competences can be developed more effectively by specifying different competence levels and by discerning subcompetences requiring targeted learning support. For example, Roorda (2010), distinguished four levels of competence in his framework of competences for sustainable development: application, integration, improvement and innovation. For an ability similar to transboundary competence, these four levels were translated to multi-disciplinary, interdisciplinary, transdisciplinary and intercultural collaboration. Roorda (2010) suggests that different academic programs may focus on different levels of competence, depending on their ambitions and possibilities. Another approach is to split a competence into its constituents. Beers (2005), for example, analysed the competence of interdisciplinary collaboration, distinguished a sequence of activities involved, and developed computer-based tools to support in particular the negotiation of common ground in multi-disciplinary (student) groups. These tools mainly serve to structure the group discussions, similar to the structured discussion boards Schoonenboom (2008) has tested in EVS. Although the effect on the quality and impact of the discussions in the EVS student groups was positive, we concluded that it did not justify the additional burden the approach placed on the group process. Applying insights from educational science and technology in competence development, generally requires curriculum and course developers to perform a balancing act between implementing micro-level improvements and safeguarding overall feasibility for the students. As indicated before, this feasibility also applies to the range of competences for sustainable development students can be expected to acquire, which is why we argued to focus on transboundary competence. In fact, we hope that in time transboundary competence will come to be considered as a generic academic competence, as broadly accepted and interwoven with domain-specific competences as the competence of critical thinking. This would mark the real integration of sustainability in higher education.

References

Barth, M. (2007) Gestaltungskompetenz durch Neue Medien? Die Rolle des Lernens mit Neuen Medien in der Bildung für eine nachhaltige Entwicklung, BWV Berliner Wissenschafts-Verlag, Berlin, Germany.

Barth, M., & Rieckmann, M. (2009). Experiencing the global dimension of sustainability: student dialogue in a European-Latin American virtual seminar. *International Journal of Development Education and Global Learning* 1(3), 22-38.

Barth, M., Godemann, J., Rieckmann, M., & Stoltenberg, U. (2007). Developing key competencies for sustainable development in higher education. *International Journal of Sustainability in Higher Education* 8(4), 416-430.

Beers, P. J. (2005). *Negotiating common ground: Tools for multidisciplinary teams.* Doctoral dissertation, Open Universiteit, The Netherlands.

Beers, P. J., Boshuizen, H. P., Kirschner, P. A., & Gijselaers, W. H. (2006). Common ground, complex problems and decision making. *Group decision and negotiation* 15 (6), 529-556.

Bromme, R. (2000). Beyond one's own perspective: The psychology of cognitive interdisciplinarity. In P. Weingart & N. Stehr (Eds.), *Practicing interdisciplinarity.* Toronto: Toronto University Press, 116- 133.

Burandt, S., & Barth, M. (2010). Learning settings to face climate change. *Journal of Cleaner Production* 18(7), 659-665.

Carlile, P. R. (2004). Transferring, translating, and transforming: An integrative framework for managing knowledge across boundaries. *Organization Science* 15(5), 555-568.

Cash, D., Clark, W. C., Alcock, F., Dickson, N., Eckley, N., & Jäger, J. (2002). Salience, credibility, legitimacy and boundaries: Linking research, assessment and decision making. Faculty Research Working Paper Series RWP02-046, John F. Kennedy School of Government, Harvard University.

Cash, D. W., Clark, W. C., Alcock, F., Dickson, N. M., Eckley, N., Guston, D. H., ... & Mitchell, R. B. (2003). Knowledge systems for sustainable development. *Proceedings of the National Academy of Sciences* 100(14), 8086-8091.

Cörvers, R., J. de Kraker (2009). Virtual campus development on the basis of subsidiarity: The EVS approach. In: M. Stansfield and T. Connolly (Eds.), *Institutional Transformation through Best Practices in Virtual Campus Development: Advancing E-Learning Policies*, pp. 179-197, Information Science Reference, Hershey/New York: IGI Global.

De Haan, G. (2006). The BLK '21'programme in Germany: a 'Gestaltungskompetenz'-based model for Education for Sustainable Development. *Environmental Education Research* 12(1), 19-32.

De Haan, G. (2010). The development of ESD-related competencies in supportive institutional frameworks. *International Review of Education* 56(2-3), 315-328.

De Kraker, J., Lansu, A. and Van Dam-Mieras, M.C. (2007). Competences and competence-based learning for sustainable development. In De Kraker, J., Lansu, A. and Van Dam-Mieras, M.C. (Eds.) *Crossing Boundaries. Innovative Learning for Sustainable Development in Higher Education*, pp. 103-114, VAS, Frankfurt a/M, Germany.

De Kraker, J. & Cörvers, R. (2014). European Virtual Seminar on Sustainable Development: international, multi-disciplinary learning in an online social network. In: Azeiteiro, U.M., Leal Filho, W., Caeiro, S., (Eds.) *E-learning and sustainability*, Peter Lang (this volume).

De Vries, B. (2013). Sustainability science. Cambridge University Press, Cambridge.

De Vries, B.J.M. and Petersen, A.C. (2009). Conceptualizing sustainable development: an assessment methodology connecting values, knowledge, worldviews and scenarios. *Ecological Economics* 68, 1006-1019.

ESF (2012) *Responses to Environmental and Societal Challenges for our Unstable Earth (RESCUE)*. ESF Forward Look –ESF-COST 'Frontier of Science' joint initiative. European Science Foundation, Strasbourg (FR) and European Cooperation in Science and Technology, Brussels (BE).

European Commission (2006). Key competences for lifelong learning – A European reference framework. Annex to the Recommendation of the European Parliament and of the Council of 18 December 2006 on key competences for lifelong learning. *Official Journal of the European Union*, 30 December 2006/L394. *http://eur-lex.europa.eu/LexUriServ/site/en/oj/2006/l_394/l_394200612 30en00100018.pdf*

European Council (2006). *Renewed EU Sustainable Development Strategy*, adopted by the European Council on 15/16 June 2006. *http://register.consilium.europa.eu/pdf/en/06/st10/st10917.en06.pdf*

EZ (2013). *Kennisprogramma Duurzaam Door: Sociale innovatie voor een groene economie (2013-2016)*. Ministerie van Economische Zaken, Den Haag.

Gallopín, G. C., Funtowicz, S., O'Connor, M., & Ravetz, J. (2001). Science for the twenty-first century: From social contract to the scientific core. *International Social Science Journal* 53(168), 219-229.

Hanning, A., Abelsson, A. P., Lundqvist, U., & Svanström, M. (2012). Are we educating engineers for sustainability? Comparison between obtained competences and Swedish industry's needs. *International Journal of Sustainability in Higher Education* 13(3), 305-320.

Janssen, M. A., & Osnas, E. E. (2005). Adaptive capacity of social-ecological systems: lessons from immune systems. *EcoHealth* 2(2), 93-101.

Jickling, B. (1992). Viewpoint: why I don't want my children to be educated for sustainable development. *The Journal of Environmental Education* 23(4), 5-8.

Johansson, M., & Læssøe, J. (2008). Mediator Competencies and Approaches to Participatory Education for Sustainable Development. Paper presented at the AERA conference sessions on ecological and environmental education SIG, New York, 24-28 March 2008.

Keulartz, F.W.J. (2005). Boundary-Work – The tension between diversity and sustainability Paper presented at the conference on Sustainability at Universities in the Czech Republic: What are Possibilities, September 2005.

Könings, K.D., Brand-Gruwel, S., van Merriënboer, J.J.G. (2005). Towards more powerful learning environments through combining the perspectives of designers, teachers and students. *British Journal of Educational Psychology* 75, 645-660.

Lans, T., Blok, V., & Wesselink, R. (2014). Learning apart and together: Towards an integrated competence framework for sustainable entrepreneurship in higher education. *Journal of Cleaner Production* 62(1), 37-47.

Leeuwis, C. (2002). Making explicit the social dimensions of cognition. In Leeuwis, C. and Pyburn, R. (Eds): *Wheelbarrows Full of Frogs – Social Learning in Natural Resource Management*, Van Gorcum, Netherlands.

Mochizuki, Y., & Fadeeva, Z. (2010). Competences for sustainable development and sustainability: significance and challenges for ESD. *International Journal of Sustainability in Higher Education* 11(4), 391-403.

Newport, D. (2008) Roundtable: Core Competencies for Sustainability Professionals, What educational background and job skills do they really need? *Sustainability: The Journal of Record* 1(4), 233-238.

Perez Salgado, F., de Kraker, J., Boon, J., & van der Klink, M. (2012). Competences for climate change education in a virtual mobility setting. *International Journal of Innovation and Sustainable Development* 6(1), 53-65.

Perez Salgado, F., Wilson, G., & van der Klink, M. (2014). Transforming academic knowledge and the concept of 'lived experience': intervention competence in an international e-learning programme. In: Azeiteiro, U.M., Leal Filho, W., Caeiro, S., (Eds.) *E-learning and sustainability*, Peter Lang (this volume).

Rieckmann, M. (2012). Future-oriented higher education: Which key competencies should be fostered through university teaching and learning? *Futures*, *44*(2), 127-135.

Roorda, N. (2010). Sailing on the winds of change. The odyssey to sustainability of the universities of applied sciences in the Netherlands. Doctoral dissertation, Maastricht University, The Netherlands.

Schoonenboom, J. (2008). The effect of a script and a structured interface in grounding discussions. *International Journal of Computer-Supported Collaborative Learning* 3(3), 327-341.

Segalàs, J., Ferrer-Balas, D., Svanström, M., Lundqvist, U., & Mulder, K. F. (2009). What has to be learnt for sustainability? A comparison of bachelor engineering education competences at three European universities. *Sustainability Science* 4(1), 17-27.

Thomas, I. (2009). Critical thinking, transformative learning, sustainable education, and problem-based learning in universities. *Journal of Transformative Education* 7(3), 245-264.

Thomas, I., Barth, M., & Day, T. (2013). Education for sustainability, graduate capabilities, professional employment: How they all connect. *Australian Journal of Environmental Education* 29(1), 33-51.

Tuinstra, W., J. de Kraker and R. Cörvers (2013). Learning to cross boundaries between science, policy and society. Proceedings of the EESD13 Conference, 22-25 September 2013, Cambridge. *http://www-eesd13.eng.cam.ac.uk/proceedings/paper98.*

UNCED (1992), *Promoting education, public awareness and training*, Agenda 21, Chapter 36, UNCED, UNESCO, Paris.

UNECE (2011). *Learning for the future: Competences in Education for Sustainable Development*, Report of the United Nations Economic Commission for Europe, Steering Committee on Education for Sustainable Development, ECE/CEP/AC.13/2011/6.

UNESCO (2005). *United Nations Decade of Education for Sustainable Development (2005-2014): International Implementation Scheme.* ED/DESD/2005/PI/01, UNESCO Education Sector, Paris. *http://unesdoc.unesco.org/images/0014/001486/148654e.pdf.*

United Nations (2012). *The Future We Want.* Outcome of the conference, Rio+20 United Nations conference on Sustainable Development, Rio de Janeiro, Brazil, 20-22 June 2012. *https://rio20.un.org/sites/rio20.un.org/files/a-conf.216l-1_english.pdf.pdf.*

Valkering, P., Beumer, C., de Kraker, J., & Ruelle, C. (2013). An analysis of learning interactions in a cross-border network for sustainable urban neighbourhood development. *Journal of Cleaner Production* 49, 85-94.

Van der Klink, M.R., & Boon, J. (2003). Competencies: the triumph of a fuzzy concept. *International Journal of Human Resources Development and Management* 3(2), 125-137.

Vare, P., & Scott, W. (2007). Learning for a change – Exploring the relationship between education and sustainable development. *Journal of Education for Sustainable Development* 1(2), 191-198.

Wals, A. E. (2010). Mirroring, gestaltswitching and transformative social learning: stepping stones for developing sustainability competence. *International Journal of Sustainability in Higher Education* 11(4), 380-390.

Wals, A. E., & Blewitt, J. (2010). Third wave sustainability in higher education: Some (inter) national trends and developments. *Green Infusions: Embedding Sustainability across the Higher Education Curriculum.* London: Earthscan, 55-74.

Wals, A. E., & Jickling, B. (2002). "Sustainability" in higher education: From doublethink and newspeak to critical thinking and meaningful learning. *International Journal of Sustainability in Higher Education* 3(3), 221-232.

Wiek, A., Bernstein, M. J., Laubichler, M., Caniglia, G., Minteer, B., & Lang, D. J. (2013). A Global Classroom for International Sustainability Education. *Creative Education* 4(4A), 19-28 .

Wiek, A., Withycombe, L., & Redman, C. L. (2011). Key competencies in sustainability: a reference framework for academic program development. *Sustainability Science* 6(2), 203-218.

Willard, M., Wiedmeyer, C., Warren Flint, R., Weedon, J. S., Woodward, R., Feldman, I., & Edwards, M. (2010). The sustainability professional: 2010 competency survey report. *Environmental Quality Management* 20(1), 49-83.

Wilson, G., Abbott, D., de Kraker, J., Perez Salgado, P., Terwisscha van Scheltinga, C. and Willems, P. (2011). The lived experience of climate change: creating open educational resources and virtual mobility for an innovative, integrative and competence-based track at masters level. *International Journal of Technology Enhanced Learning* 3(2),111-123.

Training and Employability, Competences from an e-learning undergraduate programme in Environmental Sciences

Ana Paula Martinho[1,2], **Sandra Caeiro**[1,3], **Fernando Caetano**[1,4], **Ulisses M Azeiteiro**[1,5] **and Paula Bacelar-Nicolau**[1,5]

Abstract

The trail to sustainability requires revolutionising the way environmental professionals perceive and solve environmental problems. The challenge is to prepare them to cope with societal economic and technical changes, to maintain a job and a positive role in the quest for sustainability. The environmental sector is gradually moving from an 'end-of-pipe' approach to environmental management holistic, process-based approaches, which require an entire new set of technical, social and individual skills and competences. Environmental professionals have to acknowledge the different dimensions and complexity of environmental issues, through a more proactive attitude and development of integrated solutions. In a globalisation context, environmental professionals have to develop social, ethical, creative, personal and interpersonal skills in addition to technical competences to be of value in attaining sustainability. These skills are also necessary for university environmental graduates to enter the labour market and improve their employability. This study aims to assess the development and acquisition of key skills and competences in the 1st cycle degree programme of Environmental Sciences at the Universidade Aberta, the Portuguese Distance Learning University, and their contribution to the employability of its graduates. For the assessment a questionnaire survey was conducted to the graduated students. The results are discussed within the European Tuning framework for higher education programs.

Introduction

The global economy has made largely extinct the notion of a job for life time. The imperative now is employability for life. The resulting intensification of the work process is directed at improving employee efficiency and is increasingly supported by the role of learning. The global market encourages employers to recognize and respond to changing demands. As the conditions of 'employability' change, the preparation of employable labour is expected to change (Malcolm, 2009), as well as pursuing new skills developments (McGrath and Akoojee, 2009).

---

1   Universidade Aberta, R. da Escola Politécnica, 147, Lisboa, Portugal.

2   LEAD – Laboratório de Ensino a Distância e e-learning Universidade Aberta, Portugal.

3   CENSE – Center for Environmental and Sustainability Research, FCT-UNL, Monte da Caparica, Portugal.

4   Centro de Química Estrutural, Instituto Superior Técnico, Universidade de Lisboa, Portugal.

5   CEF – Centre for Functional Ecology, Faculty of Sciences and Technology of the University of Coimbra, Portugal.

Changes in society and their impact on Higher Education are evolving ever faster. Globalisation, demographic change and rapid technological developments combine to present new challenges and opportunities for third level institutions. Future jobs are likely to require higher levels and a different mix of skills, competences and qualifications. Higher education institutions have a pivotal role to play in the success of our society and economy and their capacity to adapt to these changes and seize new opportunities is crucial (EU, 2010). Indeed these institutions have come under intense pressure to equip graduates with more than just the academic skills traditionally represented by a subject discipline and a class of degree (Mason et al., 2009; Pillai et al., 2012). There may be little to be gained from universities seeking to develop skills that are best acquired after starting employment rather than beforehand (Mason et al., 2009). According to these authors a number of reports issued by employers' associations and Higher Education institutions have urged universities to make more explicit efforts to develop the 'key, 'core', 'transferable' and/or 'generic' skills needed in many types of high-level employment. Worldwide employers complain about the lack of soft skills and employability skills among graduates as one cause for graduate unemployment (Pillai et al., 2012). These issues are particularly important in distance learning universities where most of the students belong to the working population, hoping to improve or change their jobs and where lifelong learning works as an important driving force.

Employability depends on continuous learning, being adaptable to new job demands or shifts in expertise, and the ability to acquire skills through lateral rather than upward career. Workers' employability is obtained through the acquisition of knowledge, skills, abilities, and other characteristics that are valued by current and prospective employers and thus comprises an individual's career potential (De Vos et al., 2011).

In line with the Bologna process (European Ministers of Education, 1999) the project "Tuning Education Structures in Europe" (Tuning, 2008) aimed at identifying points of reference for generic and subject-specific competences is a series of subject areas. "Competences" were defined as the description of learning outcomes: what a learner knows or is able to demonstrate after the completion of a learning process. This concerns both subject-specific competences and generic competences; generic competences can be divided into instrumental, interpersonal and systemic competences. In the Tuning project (2008), academic staff, students and employers were consulted on the generic competences they expect from graduate students and the results yielded a strong agreement with respect to the following aspects: Capacity to learn, application of knowledge to practice, capacity to adapt oneself to new situations, concern for quality, ability to work autonomously, teamwork and leadership competences, as well as research skills. Most of these competences have a clear interpersonal component (Obersat et al., 2009).

In accordance with the Tuning project, Teijeiro et al. (2013) classified the set of generic competences of graduates according to the three main domains: instrumental, interpersonal and systemic competences. Instrumental competences are defined as cognitive, methodological, technological and linguistic abilities, which are necessary for understanding, construction, operation and critical use in different professional activities. Interpersonal competences are related to one's ability to interact and network with people, as well as the ability to actively participate in specific or multidisciplinary work groups. Systemic competences are skills relative to systems, and require a combination of understanding, sensitivity and knowledge that allows one to see how the parts of a whole relate and come together.

Teijeiro et al. (2013) defends that the most relevant competences in the labour market are predominantly of the systemic type, i.e. transferable personal competences (knowledge, skills and attitudes), to the detriment of more instrumental competences related to capacities and graduate education. Nevertheless, several authors focus on identifying the most important competences for improving graduate employability (e.g. Boni and Lozano, 2007; Hennemann and Liefner, 2010; English *et al.*, 2012).

Misra and Mishra (2011) developed a conceptual framework about employability skills within business organization context. According to them employability skills can be measured based on six dimensions: skill up-gradation and career growth, task-orientation, blue-eyed boy of bosses, professional networking, and concern for time and love for challenge. On the other hand according Rothwell *et al.* (2008) employability skills are related with 'internal' factors, such as the individual's skill-set and their application to their studies, and 'external' factors such as the general state of the labour market, the strength of the university brand, and the demand for particular subject areas.

In the context of sustainable development, higher education also plays a special role in as much as it has a major influence on the way future generations will deal with the responsible societal switch points with complex requirements, which during the process of globalisation, worldwide trade, dealing with poverty as well as environment and development, have been brought upon them. Thus, society's efforts to achieve sustainability cannot be discussed without highlighting the importance of knowledge production and communication, and thus the central role of higher education institutions (Adomßent, 2013). In a globalisation context, environmental professionals have to develop new personal, interpersonal, societal and technical skills to become active throughout their professional lives and to be of value in the quest for sustainability. These professionals have to develop open mindsets, a holistic perspective of the problems and integrated solutions (Ferreira *et al.*, 2006). Most studies concerning competences and skills for employability, and their acquisition and/or development in higher education institutions, relate to face-to-face degree programmes (e.g. Tuning, 2008; Boni and Lozano, 2007; Hennemann and Liefner, 2010; Rothwell *et al.* 2008; Sirca *et al.*, 2006).

The aim of this study was to carry out, through a questionnaire survey, a self assessment of the skills and competences acquired or developed by the graduated population of an undergraduate programme in Environmental Sciences at Universidade Aberta, the Portuguese Distance learning University, and to assess if these skills and competences contribute to improve the job situation, according to the graduates' perceptions.

Case study: Undergraduate programme in Environmental Sciences

The 1st cycle degree programme in Environmental Sciences offered at Universidade Aberta (UAb) is a b-learning programme, directed to an adult public (over 21 years old), who are mostly working-students seeking professional development. The undergraduate programme in Environmental Sciences was offered for the first time in the academic year of 2007/08 following the Bologna restructuration process and is the only programme being offered mostly in an e-learning context in Portugal.

The general purpose of this undergraduate programme is to promote and develop a set of professional skills and competences within the Environmental Sciences.

The Environmental Sciences undergraduate programme integrates scientific courses in economics, earth sciences, biology and technology leading the students to the concept of sustainability science. The three years programme is composed of 30 curricular units, and totals 60 European Credit Transfer System (ECTS). The first two years (*major*) are composed of 20 compulsory curricular units in Science and Environmental Technology (40 ECTS), Biological Sciences (22 ECTS), Earth Sciences (22 ECTS), Mathematics (12 ECTS), Chemistry (12 ECTS), Physics (6 ECTS) and Legal Sciences (6 ECTS). This curricular structure provides the student with a broad-based curriculum. In the third curricular year, the student may choose one of three *minors* in an area of their preference: (i) Environmental Management and Sustainability, (ii) Natural Heritage, (iii) Environment and Health.

The Environmental Sciences undergraduate programme follows the UAb pedagogical model in its virtual class regime (Pereira *et al.*, 2007), excepting for two curricular units (Fieldwork I and Fieldwork II) which include a face-to-face component. In this learner centred pedagogical model, based on the flexibility of access, without temporal or spatial constraints, the students are responsible for knowledge building. It also relies on diversified interaction between teacher and students, students and students as well as students and learning resources in a social context.

The target public for this programme are all individuals that are concerned with environmental topics and are interested in developing their knowledge and skills in these thematic. It is also offered for technical staff from local and central governmental authorities, managers of tourism units, specially agro and sustainable tourism, managers of small enterprises involved in the production and commercialisation of biological products, employees of small private companies and public organisations that undertake environmental work, technical staff from museums, botanical gardens, natural heritage centres and local development organisations, tourism guides concerned with sustainable projects, people working in pedagogical farms and youth camps and ONG members that conduct environmental work and sustainable development.

The first graduates completed the undergraduate programme of Environmental Sciences in the academic year of 2009/2010, and the number of graduated students summed 23 in September 2013.

Methodology

This study was carried out in September 2013, when the graduate population of students from the 1st cycle degree of Environmental Sciences summed 23 graduates. All graduates were contacted using the e-mail and telephone, and inquiries (n=20) were received by e-mail. The small size of the graduate population and the methodology by which questionnaires were applied makes this a case study.

The questionnaire was composed of 9 questions, designed according to adequate criteria of clarity and objectivity. The first five questions aimed to characterise the individuals from the social-demographic point of view (age, gender and employment situation), graduation year and conclusion curricular units of the programme. The final four questions aimed to know the graduate's (i) global perception of the adequacy of competences and skills gained/developed during the programme for employability, (ii) perception of the relevance of generic competences and skills for graduate employment, (iii) perception of the level to which each competence and skill was developed during the programme, and (iv) the five more valued competences and skills for employment. A Likert scale was used for questions 4 to 8, using the categories "very high", "high", "low" and "none". For question 9, the five more valued competences/ skills for employment, the graduates were asked to ascribed a value from 5 down to 1, according to their relative "most"(=5) to "less"(=1) importance for employability. The set of generic competences and skills analysed was based on the competences of the related subject area in the Tuning Project (2008) for the degree programmes in earth and life sciences. In order to meet the goals of the study, data from the survey was processed by applying univariate statistics (frequency tables for nominal variables, adequate graphics and tables).

Results and Discussion

The small size of the population of graduates of the 1ˢᵗ cycle degree in Environmental Sciences at UAb represents a limitation to the study. This is due to the fact that UAb students are mostly working-students and take longer to complete their studies, when compared to face-to-face university programmes. Nevertheless, the high rate of return of completed questionnaires (87 %) is a good omen and enhances the obtained results of the case study.

The graduates of the 1ˢᵗ cycle degree in Environmental Sciences were half male and half female. Most graduates, 45%, were aged between 31-40 years, 20% were aged between being 41-50, 20% between 51-60, and only 15% were in the range between 21-30.

Most of these individuals graduated in 2013 (25%) and 2012 (35%), and less in the previous years (20% in 2011 and in 2010), meaning that they took between 3 and 6 years to graduate. The last curricular units (CU) completed before graduation were in most cases, two CU of 1ˢᵗ curricular year – Calculus and Physics for Environmental Sciences – followed by the CU Human Biology and other CU of the 3ʳᵈ year (figure 1).This gives an indication of the CU in which students had more difficulty throughout the programme (Calculus and Physics).

Figure 1: Last curricular units to complete before graduation (n=20).

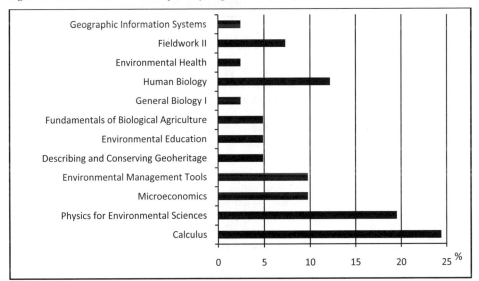

The majority of graduates of 1ˢᵗ degree programme in Environmental Sciences (80%) were currently employed (figure 2). From these, 35% were working in a position related to environmental sciences and 45% were working in a position unrelated to environmental sciences. Only 10% of the graduates were seeking employment, although 5% of these had already been in an employment situation. Still, 5% of the graduates were pursuing further studies, and 5% were neither employed nor looking for employment. The low unemployment rate among graduates of the 1ˢᵗ degree programme is related to the fact that most students of UAb have been typically working-students seeking requalification or following a study area that they are passionate about. Among graduates who answered that they were currently employed, some (15%) mentioned that they were also furthering their studies in environmental sciences at

UAb, and one mentioned he/she was awaiting requalification of employment to the environmental area, which emphasises the latter statement.

Figure 2: Employment status of graduates (n=20)

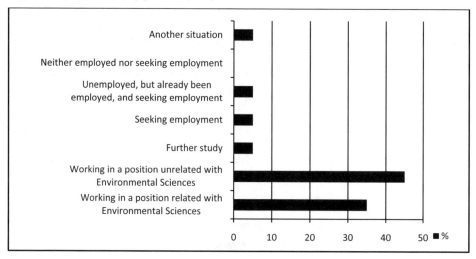

The majority (60%) of the graduates shared the opinion that the training obtained with the 1st degree programme in Environmental Sciences was adequate or very adequate (10 and 50% respectively) in terms of employment potential, while 40% of the graduates though it was neither adequate nor inadequate.

Graduate's perception of the importance of a set of 32 generic competences and skills (based on the Tuning Project, 2008) for their employment, as well as the graduate's perception of the level to which each generic competence and skill was developed during the programme are shown in table 1.

These different skills and competences can be aggregated in *Instrumental* which include items 1 to 11, 15 and 16, *Interpersonal* which include items 12, 17, 18, and 20 to 24 and *Systemic competences* including items 13, 14, 19, 25 to 29 (Vicent *et al.*, 2006; Teijeiro *et al.*, 2013), 31 and 32.

The graduate's perception of the importance of the set of generic competences and skills for their employment and their perception of the level to which each generic competence and skill was developed during the programme were globally similar, in terms of perceived relevance or development (table 1). This was the case for most of the listed *Instrumental* competences/skills (10 out of 13) and *Systemic* competences/skills (6 out of 9), and less for *Interpersonal* competences/skills (3 out of 8). Some of these skills and competences were perceived as being more developed during the 1st cycle degree programme that felt "required" for employability, particularly the *Instrumental* – Planning and time management (item 3), Structuring basic knowledge in the programme study area (item 4), Information management skills (item 11), and the Ability to integrate the interdisciplinarity between nature and society (item 32).

Some skills or competences were perceived to be less developed during the 1st degree programme Environmental Sciences than desirable for graduate's employability. These were some *Instrumental* – Capacity to apply knowledge in a practical situation (item 2), Knowledge of a second language (item 7), Basic skills in ICT (item 8) and Decision-making (item

16); some *Systemic* – Capacity to adapt to new situations (item 13), Leadership (item 19) and Initiative and entrepreneurial spirit (item 27), and mostly *Interpersonal* – Critical and self-critical abilities (item 12), Teamwork (item 17), Interpersonal skills (item 18), Ability to work in an interdisciplinary team or international context (items 20 and 23) and Understanding of cultures and customs of other countries (item 24).

These results are a direct consequence of the fact that the pedagogical offer of UAb at the 1st cycle level has given special emphasis to the expansion of Portuguese language and culture within the Lusophony space, and is hence presented in Portuguese language (unlike the 2nd cycle degrees) and also to the characteristics of the virtual pedagogical model for 1st cycle degrees of UAb which relies on individual student work (unlike the 2nd cycle in which student work is fundamentally of collaborative nature). Hence, the Instrumental competences related to the domain of a 2nd language and the set of Interpersonal competences herein analysed, have not been a preferential target for development during 1st cycle degrees, which is well reflected in the results of this study. Also some of the Systemic competences and skills which were perceived as less developed than required for employability, such as Leadership (item 19) and Initiative and entrepreneurial spirit (item 27) are not though as central to a 1st degree cycle in Environmental Sciences, and have not been targeted for development in the current curriculum.

The five most important skills and competences for employability chosen by the graduates from the list of 32 are presented in figure 3. The top valued skills/competences for employability (weighted sum over 15) were the Capacity to analyse and synthesise (item 1), Capacity to apply knowledge in a practical situation (item 2), Teamwork (item 17) and Ability to work autonomously (item 25). Just below these, the following set of competences / skills were valued: Information management skills (item 11), Problem solving (item 15), Decision-making (item 16), Ability to work in an interdisciplinary team (item 20), Initiative and entrepreneurial spirit (item 27), Concern for quality (item 29) and Ability to integrate interdisciplinarity between nature and society (i.e. environmental, social and economic) (item 32).

Table 1: *Student perception of the relevance of each skill and competence for employability, and of their level of development during the 1st cycle degree programme in Environmental Sciences (grey shadow indicates the most selected option by graduates) (n=20)*

Skills / Competences	Relevance for employability (%)				Development during degree programme (%)			
	3 = very high	2 = high	1 = low	0 = none	3 = very high	2 = high	1 = low	0 = none
1. Capacity to analyse and synthesise	80	20	0	0	60	40	0	0
2. Capacity to apply knowledge in a practical situation	80	20	0	0	35	55	10	0
3. Planning and time management	70	30	0	0	90	10	0	0
4. Structuring basic knowledge in the programme study area	50	40	5	0	55	35	0	0
5. Structuring basic knowledge in professional activity	35	45	10	5	25	55	10	5
6. Oral and written communication in native language	40	55	5	0	30	60	5	0
7. Knowledge of a second language	5	65	20	10	0	30	50	20

Martinho, Caeiro, Caetano, Azeiteiro and Bacelar-Nicolau

Skills / Competences	Relevance for employability (%)				Development during degree programme (%)			
	3 = very high	2 = high	1 = low	0 = none	3 = very high	2 = high	1 = low	0 = none
8. Basic skills in ICT	20	75	5	0	25	55	20	0
9. Research skills	55	45	0	0	70	20	10	0
10. Learning ability	75	25	0	0	80	20	0	0
11. Information management skills	40	60	0	0	65	35	0	0
12. Critical and self-critical abilities	65	35	0	0	40	55	5	0
13. Capacity to adapt to new situations	60	40	0	0	50	40	10	0
14. Capacity to generate new ideas (creativity)	65	35	0	0	30	65	5	0
15. Problem solving	55	45	0	0	40	60	0	0
16. Decision-making	75	25	0	0	30	60	10	0
17. Teamwork	50	45	5	0	45	25	25	5
18. Interpersonal skills	45	50	5	0	40	45	15	0
19. Leadership	35	55	5	5	5	50	25	20
20. Ability to work in an interdisciplinary team	45	50	5	0	40	45	15	0
21. Ability to communicate with non-experts	45	45	10	0	40	45	15	0
22. Appreciation of diversity and multiculturality	45	55	0	0	55	40	0	5
23. Ability to work in an international context	10	75	15	5	10	45	25	15
24. Understanding of cultures and customs of other countries	10	55	35	0	5	45	35	15
25. Ability to work autonomously	60	40	0	0	60	40	0	0
26. Project design and management	10	65	20	5	0	75	20	5
27. Initiative and entrepreneurial spirit	35	60	5	0	15	60	25	0
28. Ethical commitment	75	25	0	0	75	25	0	0
29. Concern for quality	85	15	0	0	80	20	0	0
30. Will to succeed	85	15	0	0	80	20	0	0
31. Ability to integrate the interdisciplinary science of sustainability	85	15	0	0	75	20	5	
32. Ability to integrate interdisciplinarity between nature and society (i.e. environmental, social and economic)	65	30	5	0	75	25	0	0

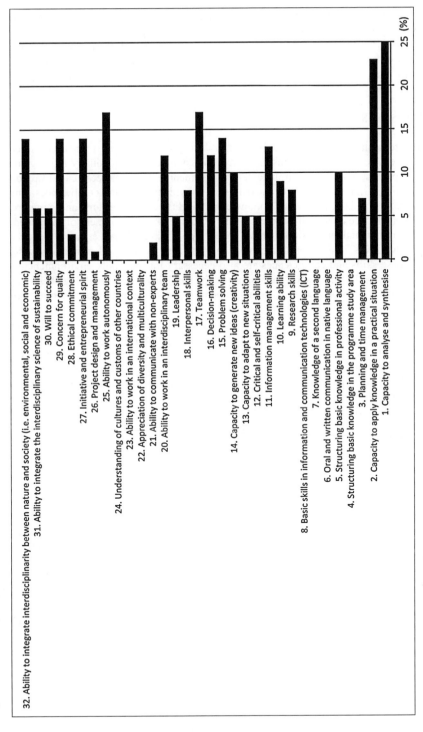

Figure 3: Graduates' perception of the 5 most important generic competences and skills for employability

(for each competence/skill the weighed sum was calculated attributing a value from 5 down to 1, according to their relative importance to employability, from "most" to "least" important) (n=20)

It is interesting to note that from the set of competences and skills rated 'very important' for employability by the graduates (table 1) only a small group were chosen as the five most relevant competences for employability and not all of these match the previous choices. This group of five most relevant competences and skills do, however, mostly correspond to those pointed by other studies (e.g. Tuning, 2008; Sirca *et al.*, 2006; Pillai *et al.*, 2012)

Conclusions

The path to sustainability requires changing the way environmental professionals perceive and solve environmental problems. The environmental sector is gradually moving from an 'end-of-pipe' approach to environmental management holistic, process-based approaches, which require a new set of technical, social and individual skills and competences. Environmental professionals have thus to acknowledge the different dimensions and complexity of environmental issues, through a more proactive attitude and development of integrated solutions. Environmental professionals have to develop social, ethical, creative, personal and interpersonal skills, in addition to technical competences, to be of value in attaining sustainability. These skills and competences are also necessary for university environmental graduates to enter the labour market and improve their employability, enabling their mobility, competitiveness and consequently, giving impetus to the development of green jobs and the development of society and economy.

Essential to the labour markets and its changing role, within the employability issues, is the role of learning. Lifelong learning has been moved from a peripheral place to a central feature of European Government economic policy. Hence, industry is encouraged to adopt new cultural attitudes and practices by locating learning at the centre of workers' lives. The role for 'learning' is to help workers fitting the needs of industry by becoming work-centred, efficient and compliant employees. As the unspecified needs of industry change so, with the aid of lifelong learning, labour is required to develop or adapt its skills.

This study indicates that the 1st cycle e-learning programme in Environmental Sciences of Universidade Aberta, is globally adjusted to the employability requirements, through the acquisition and/or development of competences and skills of its current graduates. However, some relevant competences / skills, particularly Interpersonal competences and a few central Instrumental competences e.g. related to the application of knowledge to practical situations, were not sufficiently developed.

The 1st cycle e-learning programme in Environmental Sciences needs, therefore, to be more closely aligned with the needs of the labour market, e.g. by introducing changes in the university curricula, pedagogical tools or technological tools designed to potentiate the development of key competences and skills for employability in its area of expertise. The subject of Environmental Sciences delivers many opportunities to use innovative and problem-based teaching methods. Topics like environmental protection, planning and sustainable development, expose students to practical relevant issues. The systematic monitoring of the graduate career requirements is a topic that needs sustained research as it grants insight into the utility of current curricula, in an ever faster changing labour market.

References

Adomßent, M. (2013). Exploring universities' transformative potential for sustainability-bound learning in changing landscapes of knowledge communication. *Journal of Cleaner Production* 49, 11-24.

Boni, A. and Lozano, J. F. (2007). The generic competences: an opportunity for ethical learning in the European convergence in higher education. *High Education*, 54: 819-831.

De Vos, A., Hauw, S. and Van der Heijden, B. (2011). Competency development and career success: The mediating role of employability, *Journal of Vocational Behavior* 79, 438-447.

European Ministers of Education (1999) Joint Declaration of the European Ministers of Education, Bologna, 19th June 1999. Retrieved December 15, 2013 from *http://europa. eu/legislation_summaries/education_training_youth/lifelong_learning/c11088_en*.

EU (2010). The EU contribution to the European Higher Education Area. European Commission. Luxembourg. ISBN: 978-92-79-15103-3. Retrieved December 2, 2013 from *http://bookshop.europa.eu/en/the-eu-contribution-to-the-european-higher-education-area-pbNC8010233/*.

English, D., Manton, E., Sami, A. R. and Dubey, A. (2012). A comparison of the views of college of business graduate and undergraduate students on qualities needed in the workplace. *College Student Journal* 46 (2), 427-435.

Ferreira, A. J. D., Lopes, M. A. R. and Morais, J. P. F. (2006). Environmental management and audit schemes implementation as an educational tool for sustainability. *Journal of Cleaner Production* 14, 973-982.

Hennemann, S. and Liefner, I. (2010). Employability of German geography graduates: the mismatch between knowledge acquired and competences required. *Journal of Geography in Higher Education* 34(2), 215-230.

Malcolm, B. (2009). Learning, labour and employability. *Studies in the Education of Adults* 41(1), 39-52.

Mason, G., Williams, G. and Cranmer, S. (2009). Employability skills initiatives in higher education: what effects do they have on graduate labour market outcomes? *Education Economics* 17(1), 1-30.

McGrath, S. and Akoojee, S. (2009). Vocational education and training for sustainability in South Africa: The role of public and private provision. *International Journal of Educational Development* 29, 149-156.

Misra, R. K. and Mishra, P. (2011). Employability Skills: The Conceptual Framework & Scale Development. *The Indian Journal of Industrial Relations* 46(4), 650-660.

Obersat, U., Gallifa, J., Farriols, N. and Vilaregut, A. (2009). Training Emotional and Social Competences in Higher Education: The Seminar Methodology. *Higher Education in Europe* 34 (3-4), 523-533.

Pereira, A., Mendes, A. Q., Morgado, L., Amante, L. and Bidarra, J. (2008). Universidade Aberta's Pedagogical Model for distance education. Universidade Aberta. Retrieved December 20, 2013 from *https://repositorioaberto.uab.pt/bitstream/10400.2/2388/1/MPV_uaberta_english.pdf*.

Pillai, S., Khan, M., Ibrahim, I. S. and Raphael, S. (2012). Enhancing employability through industrial training in the Malaysian context. *High Education* 63, 187-204.

Rothwell, A., Herbert, I. and Rothwell, F. (2008). Self-perceived employability: Construction and initial validation of a scale for university students. *Journal of Vocational Behavior* 73, 1-12.

Sirca, N. T., Nastav, B., Lesjak, D. and Sulcic, V. (2006). The Labour Market, Graduate Competences and Study Programme Development: A Case Study. *Higher Education in Europe* 31(1), 53-64.

Teijeiro, M., Rungo, P. and Freire, M. J. (2013). Graduate competencies and employability: The impact of matching firms' needs and personal attainments. *Economics of Education Review* 34, 286-295.

Tuning (2008)."Tuning Education Structures in Europe" Universities' contribution to the Bologna Process. 2nd ed. Retrieved December 20, 2013 from *http://www.unideusto.org/ tuningeu/images/ stories/Publications/ENGLISH_BROCHURE_FOR_WEBSITE.pdf.*

Vicent, L., Ávila, X., Riera, J., Badia, D., Anguera, J. and Montero, J. (2006). Appropriateness of e-learning resources for the development of transversal skills in the new European Higher Education Area. 36th ASEE/IEEE Frontiers in Education Conference S2J-6, October 28-31, San Diego, CA. Retrieved December 20, 2013 from *http://ieeexplore.ieee.org/ stamp/stamp.jsp?tp= &arnumber=4117111.*

Transforming academic knowledge and the concept of Lived Experience: Intervention Competence in an international e-learning programme

Francisca Pérez Salgado[1], Gordon Wilson[2] and Marcel van der Klink[3]

Abstract

Within sustainability issues climate change is recognised as one of the most challenging and defining for our future. However, the learning and teaching in this field is perceived by students as complex and contradictory, and it leaves them with uncertainties with respect to their professional practice. This chapter describes a solution to flaws observed in university programmes.

The concept of the Lived Experience explains the existence of several perspectives at the same time. It connects abstract and distant scientific knowledge with personal, local and cultural diversity. It treats epistemological diversity as a resource for social learning and holistic knowledge. The authors consider this concept to be important and perhaps even crucial for the domain of sustainability, where it can be used to expand knowledge and linking academia with professionals and citizens.

In an open access Masters track called the 'Lived Experience of Climate Change', the learning goals and outcomes are operationalised using the concept of 'competence'. Complementing Transboundary Competence, this paper focuses on Intervention Competence. Intervention Competence combines strategic-political thinking with personal goal-directedness, formulating solutions and actions for climate change issues, in awareness of societal aspects. Thus the student's ability to transform academic knowledge to sustainable solutions is developed.

By adding Intervention Competence to university programmes, students are encouraged to engage with each other and their teachers to propose realistic and sustainable solutions to sustainability challenges. They use their diversity as a resource, a process that may be enhanced by virtual mobility arrangements between several universities.

---

1   Faculty of Management, Science and Technology, Open University of the Netherlands, P.O.Box 2960, 6401 DL Heerlen, The Netherlands.
2   Faculty of Mathematics, Computing and Technology, The Open University, Walton Hall, Milton Keynes, MK7 6AA, United Kingdom.
3   Welten Institute, Open University of the Netherlands, P.O.Box 2960, 6401 DL Heerlen, The Netherlands.

Introduction

Sustainable development as a process of social learning and action

The United Nations Decade of Education for Sustainable Development (UNDESD, 2005-2014) is drawing to a close as this chapter is being written. It is pertinent to ask the questions: 'What is its legacy? What should we continue to promote in future years?'

This chapter explores a particular aspect of the legacy which we consider to be of fundamental importance: a conceptual shift from the idea of sustainable development as a scientifically definable and agreed end point for society once all the relevant facts are known, to a process of social learning and action. The word 'process' in the formulation signals the idea of continual adjustment and occasionally major shifts in practice, while 'social learning' embodies the bringing together of different perspectives in dialogue and debate to create new knowledge that informs and makes possible human interventions for sustainability. The phrase 'social learning and action', which Kolb (Kolb, 1984) summarises as 'experiential learning', conveys the idea of a cyclical relationship between the two, which is generally known as the action-learning cycle (ibid.). Action-learning in turn can be viewed as the mechanism for the aforementioned process of sustainable development. However, as Jarvis (2012) demonstrated, not every experience functions as a driving force that generates high quality learning, since often the potential for learning is overlooked or avoidance of learning opportunities appear, for example in the case of low levels of self-efficacy of the participants involved (Bandura, 1997).

Fundamental philosophical principles behind this shift in how we understand sustainable development include:

* Constructivist views of knowledge and scientific uncertainty. More recently, how to work with scientific uncertainty in complex societal challenges has become the subject of what is termed 'post-normal science' (Funtowicz and Ravetz, 1993);
* Climate change as a prime example of this way of knowledge construction. Arguably all scientific knowledge obtained outside of strictly controlled laboratory conditions is less precise, and subject to modelling where the modeller's assumptions inevitably introduce incomplete, subjective dimensions and uncertainties;
* Difference as a primary source for constructing new knowledge, which draws on post-modernist notions of the validity of different perspectives, which is acknowledged in emerging theories on knowledge productivity and co-creation (see, for example, Chan and Dixon, 2012; Voorberg, Bekker and Tummers, 2013);
* The fundamental human ability to reflect and engage with others to generate new knowledge, what Habermas (Habermas, 1990, 2011) calls 'communicative action' and which is echoed in leading adult learning theories, such as the concept of reflective practitioner that was proposed by Schön (1987).

This conceptual shift has added a dynamic element to the classic, and still most cited, definition of sustainable development, that of the 1987 World Commission for Environment and Development chaired by the then Norwegian Prime Minister, Gro Harlem Brundtland:

> '... development that meets the needs of the present without compromising the ability of future generations to meet their own needs' (Brundtland,1987).

As with the Brundtland definition, however, the UNDESD-inspired change towards 'sustainable-development-as-learning-process' is abstract and requires grounding in social reality where difference also reflects relations of power and inequality, and different perspectives do not in practice have equal validity. These power relations are further reflected in human engagement with the result that emerging new knowledge reflects the interests of the most powerful (Foucault, 1980; Haraway, 1988, 1989). Moreover, engagement is more a process of contestation than reasoned dialogue and debate (Hulme, 2009).

The outline of this chapter is as follows: in section 2 we introduce the concept of Lived Experience as a powerful way of crystallising and taking further the shift in thinking about Sustainable Development as outlined above. In section 3 we focus on the explanatory capacity of Lived Experience and its actual implementation in a European e-learning Master's programme on Climate Change. In section 4 we advocate a competence-based approach to education for sustainable development as such an approach is especially useful for integration of different knowledge domains and skills. In section 5 we focus specifically on Intervention Competence in relation to Sustainable Development given that the process of intervention in this domain is a major challenge when faced by multiple perspectives that derive from both science and lived experiences. We end in section 6 with some conclusions, which include our advocacy of Lived Experience and Intervention Competence in open access e-curricula, and suggestions for future research.

Sustainable Development and the concept of Lived Experience

The above ideas have led to the introduction of a people-centred concept, the 'Lived Experience of Climate Change' (Abbott and Wilson, 2012; Wilson et al., 2011), which was exemplified in an e-learning Master's programme by a diverse group of researchers in a European Union Erasmus project (LECH-e, 2009-2012; see website mentioned in Wilson, 2011). Lived Experience is knowledge gained by people over time through engagement with each other and learning from actions. It is thus an evolving knowledge, and hence a process, but it is also influenced by more enduring factors such as social class, gender, ethnicity and local cultural values. The concept takes on board the overlapping notions of social, experiential and situated learning, and action-learning cycles, while also recognising the social conditions of knowledge production and engagement (Johnson and Wilson, 2009, p. 128). It is a powerful concept because it focuses on people, and as such (Abbott and Wilson, 2013):

- Can be related to how people interpret differently the same global, societal challenge, such as climate change. It explains the variety of simultaneous, co-existing and often contradictory perspectives on this and other challenges;
- Recognises the social conditions of human actions as well as those of knowledge production;
- Reclaims everyday experiential knowledge as an important factor in interpretation of global challenges. In other words, science is not the only truth that informs global challenges such as climate change;
- Potentially overcomes the science-citizen dichotomy, because scientists are also human beings with lived experiences that filter their scientific findings into knowledge and actions;
- Leads to new ways of examining public engagement with science and the use of knowledge in political policy making. These new ways take us beyond a knowledge-deficit model (*'If only the public were communicated the scientific facts, they would recognise the importance of climate change'*), to one of engagement between different forms of knowledge to create new knowledge.

Within sustainable development, climate change is recognised as one of the most challenging and defining issues for our future. Equally, therefore, our concept of the Lived Experience of climate change may be extended and used in relation to the broader topic of sustainable development.

The Erasmus project concerned itself with introducing and examining the concept as complementary to traditional scientific (physical and social) ways of approaching education in sustainable development. It involved a truly interdisciplinary team of natural and social scientists, and engineers. The project also made a preliminary attempt to introduce and operationalise through its virtual mobility platform the competences that are needed to work with real-life, transboundary challenges of climate change policy (Chapter 2 in this volume) and intervention strategies. These challenges are associated with knowledge boundaries within and between the physical and social sciences, and how they are interpreted through lived experiences.

A diversity of perspectives, strengthened by e-learning

To the extent that it develops the idea of social learning and action, the concept of Lived Experience acknowledges the UNDESD. This is not a one-way process, however and in order to close the loop we have to ask a further question, which is: 'What does Lived Experience mean for education for sustainable development?'

We identify three dimensions:

a) Lived Experience is important for education for sustainable development because it has explanatory power. As stated above, the concept of Lived Experience, both explains the presence and validity of many competing perspectives at the same time;
b) Accepting the explanatory power of Lived Experience is to accept the validity of multiple perspectives on sustainable development. This then raises the further challenge of how to work *with* such diversity rather than *against* it. Our starting point here is the constructivist approach to knowledge where diversity is fundamental. We construct new knowledge through engagement with our differences – both big and small – not through being the same;
c) E-learning has a vital role to play in both using Lived Experience as an explanatory feature of education for sustainable development, and for developing transboundary and intervention competences. This is because of the potential of e-learning to provide quality education en masse and across geographical boundaries, where the sheer numbers and expanse across our earth ensure a rich diversity of perspectives. In this endeavour, students are at least as much creators of new knowledge out of their diversity as are the teachers. Thus, not only do the following sections focus on Intervention Competence, they do so in the context of e-learning.

The explanatory power of Lived Experience in education for sustainable development

It is not surprising that many students are to some extent bewildered when embarking on a course related to sustainable development. The learning and teaching in this field is perceived by them as complex and contradictory, and it leaves them with uncertainties with respect to their professional practice. The results of a world survey among many (more than 1000) students from several countries worldwide indicate that students desire a better coverage of climate change education in their university programmes (Leal Filho, 2010; tables 5 and 7, p. 12 and p. 16).

One reason for such bewilderment is that their courses are often partial, focusing on one or a few aspects. Although there is a broad consensus on the fact that acquiring some basic knowledge of meteorological, geo-physical, geo-chemical aspects is necessary, it is now broadly agreed that this is in itself insufficient in an educational programme on human-induced climate change. The climate change issue has a complexity that requires additional knowledge and skills besides the natural sciences.

Consensus has grown that climate change education, just as education for sustainable development, should include the following aspects in its programme (UNESCO, 2004, 2010):

- raising awareness of the different levels of, and perspectives on, human-induced environmental problems and challenges (different temporal and spatial scales; economic, political, societal and cultural diversity);
- taking an interdisciplinary approach.

Moreover, and a second reason for potential bewilderment is that even within each aspect there are no definitive conclusions. Thus the physical science is uncertain and contested, while the economic and social implications discussed by social science approaches are even more so. Management and intervention, moreover, always appear difficult and 'political', where searching for prescriptions is futile. In short, science, social science and management education does not and cannot provide definitive answers to the questions of sustainable development and to expect them to do so is a recipe for disappointment. Nowhere is this better exemplified than in the many, competing perspectives on climate change. This is not simply a debate between those who see overwhelming evidence for human-induced global warming and those who deny such evidence. Even among those who are not deniers of global warming, there is much dispute over the degree and rate of temperature rise, and its effect on climate and subsequent impact on life.

With the concept of the Lived Experience (of Climate Change or any other subject related to sustainable development), however, one can teach students the existence of several perspectives at the same time. More importantly perhaps, one can use the concept to *explain why* these multiple perspectives exist, not only among the academic disciplines, but within them, and also why the general public is a key actor. In short it provides an organising idea for coherence of what is often seemingly incoherent. This needs, however, a rethinking of the pedagogy: what is an appropriate way to teach this?

Competence-based education in an e-learning context

Unleashing the power of e-learning demands a specific pedagogy, because without a well-thought pedagogy e-learning usually does not result into effective or efficient education.

The concept of 'competence' is increasingly being adopted in higher education and life-long-learning, especially in fields where an integration of different knowledge domains and skills is desirable, such as sustainable development and climate change. Competences and a competence-based curriculum are therefore at the heart of the curriculum development. A crucial question follows: how many competences are needed, are there key competences and how can tasks be designed in a both meaningful but also 'reliable' way?

It is important to be explicit about which definition one uses, since different cultural foci and learning theories lead to (implicit) different definitions (Boon and Van der Klink, 2003; Pérez Salgado, De Kraker, Boon and Van der Klink, 2012) and contribute to misunderstandings.

In the e-learning 'Lived Experience of Climate Change'-programme competence is defined as a 'cluster of skills and knowledge which can be learned through tasks performed in the workplace or through high-fidelity simulations of authentic work environments. By choosing this approach one is able to make a relatively complete description of a competence. Both for the communication to students and for a reliable assessment, a comprehensive and thorough definition is crucial and highly desirable.

Within competence-based curricula Van der Klink, Schlusmans and Boon (2007) picture two types which will be outlined hereafter and are compared in table 1. The first curriculum is defined as mainstream competence-based curriculum (MCC) and this came into existence as a reaction to traditional curricula that were mainly knowledge-oriented and were insufficiently focused on the labour market needs. The main purpose of the MCC is to improve the match between curriculum and labour market demands by offering a curriculum that is mainly multi-disciplinary, learner-centred and based upon authentic situations. The locus of control is still the educational provider, i.e. the university.

One of the pitfalls of a MCC is that the present needs of employers are perhaps too dominantly incorporated in the curriculum, which is especially problematic in domains in which innovation and change are the constant factors combined with ambiguity regarding the future developments of a vocation or profession. For that reason Van der Klink et al. (2007) propose an alternative view on competence-based education which allows students themselves to steer much more the content of their learning: the self-directed competence-based curriculum (SDCC).

Table 1:   Key features of mainstream competence-based and self-directed competence-based curricula

	Mainstream competence-based curricula	**Self-directed competence-based curricula**
Main purposes	– to achieve a better match between curricula and labour market demands – to give out degrees and certificates	– knowledge co-production and empowerment of the individual learner – to give out degrees and certificates
Content of the curriculum	Fixed curriculum based on authentic situations, cases in which students have to demonstrate broad competences, mainly multidisciplinary	No central curriculum but a personal learning plan in which the student decides which learning situations he or she will use to acquire competences
Student activity	Carrying out tasks, solving problems in kinds of project-based learning activities	Students themselves decide which activities they undertake to acquire the competences
Teaching style	Combination of teacher- and student-centred learning. Activities to be decided mainly by the student	Combination of teacher- and student-centred learning. Activities and learning goals to be decided mainly by the student
Flexibility within the curriculum	The curriculum is adapted to the entry-level and personal needs of the student within the framework decided upon by the university	The student's personal preferences and needs are central.
Assessment	Demonstrating competences Summative and formative assessments. Emphasis on types of performance assessments, also in authentic situations (e.g. workplace)	Demonstrating competences in a way that is decided by the student. This applies to summative and formative assessments as well. Different types of assessment, including portfolio assessment

based on Van der Klink et al., 2007

Table 1 compares both types of competence-based educational curricula. It goes without saying that the SDCC appears to be slightly provocative and perhaps its full implementation is a harsh endeavor within the university context. However, education for sustainable development leads to either MCC or SDCC, but increasingly SDCC seems a better alternative for the challenges posed. In a SDCC-curriculum the concept of Lived Experience can be fully embraced, since its societal context is much broader and diverse than only the labour market.

Intervention Competence for Sustainable Development

As explained in the previous section, the use of competences in education for sustainable development can be appropriate, as long as one uses clear definitions. However, which are the key competences, and how many should be learned by students?

Here, we identify two key competences: transboundary and intervention competence. Since the e-learning programme is on Climate Change, we mention that as the field of application here. However, the competences can be easily broadened and used in other fields of sustainable development.

A key competence is to learn to think, collaborate, and communicate across the boundaries of the different perspectives. This ability for communicative engagement across boundaries is referred to as Transboundary Competence (de Kraker, Lansu, & van Dam-Mieras, 2007; also Chapter 2 of this volume). Another key competence is to be able to critically discuss how relevant scientific and experiential knowledge can inform solution(s) to the societal problems to which climate change exacerbates (or at least contributes) and, as a following step, to be able to reach decisions concerning intervention strategies. This ability is called Intervention Competence (Pérez Salgado et al., 2012) and is the focus for the remainder of this chapter.

With this competence students learn to make the step from studying a problem to formulating ways and options to reaching decisions or to interventions. A more formal definition of the Intervention Competence for Climate Change education would be:

> *'the ability to devise or propose, independently and after consultation with relevan actors, one or several sustainable solution(s) or to reach decisions for a climate change problem and indicate its consequences for the biophysical and socio- cultural environment.'*

Intervention Competence combines the scientific domains and skills, and experiential knowledge, to create an 'integrated' assessment, from which decisions can be reached and interventions designed. Firstly, it requires insight into the natural scientific (geo-bio-chemical) knowledge. Secondly, it requires social scientific knowledge of the social, economic, political, gender and cultural dimensions of climate change. Thirdly, it requires direct engagement with actual lived experiences which represent how people think and feel about a phenomenon, and hence are complementary to natural and social scientific knowledge. Thus, Intervention Competence focuses on the 'problem-solving' or 'decision-making' aspect. It leads to knowledge that is co-produced with a range of societal actors and uses this knowledge for reaching widely acceptable decisions concerning appropriate interventions and their design. Thus it can serve social and societal change.

Whereas Transboundary Competence concerns the ability to engage productively using different perspectives on sustainable development and communicating productively with different groups (cultural, gender, class, racial), with Intervention Competence we take this as the starting point for further development.

This competence development involves:

- appreciating the importance of (trying to) reach to decisions or interventions;
- being aware of a *multitude of solutions*, related to different perspectives and to different groups of actors;
- being able to translate this diversity into propositions and decisions for interventions (actions);
- being able to engage in political-strategic thinking, combined with personal and individual goal-directedness (strategic decision making);
- *being able to steer towards collectively produced proposals and decisions*, articulating policies and/or proposing initiatives which challenge the existing non-sustainable practices, and are change-effective.

It goes without saying that traditional straight-forward intervention strategies are not included in the notion of the Intervention Competence as described here. It has no relation to Roger's well-known innovation model (Roger, 1995), since this prescribes a top-down change approach in innovation processes. Emerging approaches in the area of knowledge production (Stam, 2007), innovation and social capital (Nahapiet and Goshal, 1998) appear to be more in correspondence with the views on the Intervention Competence that are expressed in this chapter. However, emerging approaches are still in their infancy and usually lack a proposal for an elaborated strategy that outlines the different steps and activities to be taken in the process of intervening (Van der Klink, 2012).

Exploring sophisticated approaches in other professional domains might be supportive in further rethinking the content of Intervention Competence. Here we would like to point at Intervention Mapping as a promising approach that is often applied in different healthcare contexts to tackle complex health issues in an evidence-based manner and focuses on the change of behaviour of the targeted users of the intervention (see for examples Michie et al., 2008; Wolfers et al., 2007).

The e-learning programme LECH-e introduced Intervention Competence (Pérez Salgado et al., 2012), but the focus was mainly on the development of Transboundary Competence. Intervention Competence needs to be further developed both conceptually and didactically; evaluations are needed as well. Through designing competence-tasks for students in the e-learning environment in such a way that they may practise intervention competence at each level, students can gain insight and command step-by-step.

When students master Intervention Competence they are prepared for their future roles as professionals in the field and as active citizens. In fact, Intervention Competence can be seen as the lynchpin between science and scientific knowledge at a university on the one hand and change processes in society and personal action(s) on the other. In addition, the combination of Lived Experience and Intervention Competence allows students, professionals and citizens to link their own experience to science and to (often) remote government policies. As such, they learn how to appreciate the diversity of different Lived Experiences, and accept and work on a multitude of different 'best solutions'.

Conclusions

E-learning harbours innovative examples in the field of education for sustainable development. In this chapter, we have explained and explored further two powerful concepts: Lived Experience and Intervention Competence. Both were introduced and developed by a group of

European researchers in an open access Masters Track 'The Lived Experience of Climate Change'.

Traditional e-learning programmes focus mainly on disciplinary knowledge reproduction, whereas education for sustainable development requires an integration of academic fields and an appreciation of societal aspects.

The people-centred concept of Lived Experience is integrative, pluralistic and holistic, and is self-explanatory with respect to a diversity of perspectives on sustainable development, since lived experience will by its nature vary. The starting point is a constructivist approach to knowledge where diversity is fundamental. New knowledge is constructed through engagement with our differences – both big and small – not through being the same. In addition to individual variability, the concept embodies more enduring societal perspectives (race, class, gender, culture). It connects abstract scientific knowledge to local, personal and cultural diversity, and thus explains a diversity of perspectives, and in addition, allows for a diversity of interventions. In this respect, it is an example of social learning and gears towards an action competence-based learning process. In this process, new knowledge is constructed.

We introduce a (partly) self-directed competence-based curriculum, in which the concept of Lived Experience is developed using two competences, which we consider to be crucial for Sustainable Development. Both focus on understanding, managing and working with diversity: Transboundary and Intervention Competence. Transboundary Competence concerns transcending the knowledge boundaries associated with multiple perspectives to arrive at new knowledge (see chapter 2). Intervention Competence concerns arriving at decisions and designing appropriate interventions for sustainable development from the new knowledge so derived. In this chapter we develop this competence further pedagogically.

Through learning and training with respect to both Lived Experience and Intervention Competence in open access e-curricula, students and citizens all over the world may overcome their bewilderment with respect to sustainable development, better grasp its complexity, and envisage and work on effective solutions.

We stress, however, that this is work in progress and we are still near the start of a long journey. Our argument above sparks two immediate questions for future research around education for sustainable development:

1. How can the explanatory power of Lived Experience be operationalised so that students (and professionals) can then carry it forward to their future professional lives, and as citizens?
2. Moving beyond the general, what are the specific dynamics of Intervention Competences that enable them to facilitate effectively what they are supposed to facilitate, namely intervention in multi-actor settings?

These questions are currently the basis of further research, in which students, professionals and citizens are being involved.

References

Abbott, D. and Wilson, G. (2014). "Climate change: lived experience, policy and public action" *International Journal of Climate Change: Strategies and Management* 6(1), 5-18.

Abbott, D. and Wilson, G. (2012). "The Lived Experience of Climate Change: Complementing the Natural and Social Sciences for Knowledge, Policy and Action", *International Journal of Climate Change: impacts and Responses* 3(4), 99-114.

Bandura, A. (1997). *Self-efficacy in changing societies.* Cambridge: Cambridge University Press.

Boon, M. J., & Klink, M. R. van der (2003). Competence: The triumph of a fuzzy concept. *International Journal of Human Resources Development & Management* 3(2), 56-77.

Brundtland, G. (ed.) (1987). *Our common future: The world commission on environment and development*, Oxford, Oxford University Press (Commonly known as the Brundtland Report).

Chan, A. & Dixon, C. (2012). Post-modern perspectives on organisational learning. *International Journal of Learning and Intellectual Capital* 9(1-2),137-150.

De Kraker, J., Lansu, A., & van Dam-Mieras, M. C. E. (2007). Competences and competence-based learning for sustainable development. In J. de Kraker, A. Lansu, & M. C. E. van Dam-Mieras (Eds.), *Crossing Boundaries. Innovative learning for Sustainable Development in Higher Education* (pp. 25-36). Frankfurt: VAS.

De Kraker, J., Cörvers, R., & Lansu, A. (2014). E-learning for sustainable development: linking virtual mobility and transboundary competence development, *chapter 3 in this volume.*

Foucault, M. (1980). "Truth and power", in Gordon, C (ed.) Michel Foucault: Power/Knowledge: Selected Interviews and Other Writings 1972-1977, The Harvester Press, Pearson Education Ltd, London.

Funtowicz, S.O. and Ravetz, J.R. (1993). The Emergence of Post-Normal Science, in: René von Schomberg, (ed.), Science, Politics and Morality. Scientific Uncertainty and Decision Making 85-123.

Funtowicz, S. and Ravetz, J. (ND) "Post-normal science: environmental policy under conditions of complexity", NUSAP.net. Retrieved from: *http://www.nusap.net/sections.php? op=viewarticle &artid=13.*

Habermas, J. (1990). *Moral Consciousness and Communicative Action*, Polity Press, Cambridge.

Habermas, J. (2011). *"Zur Verfassung Europas – Ein Essay"*, Suhrkamp Verlag Berlin.

Haraway, D. (1988). Situated knowledges: The Science Question in Feminism and the Privilege of Partial Perspective. *Feminist Studies*, 14(3), 575-599.

Haraway, D. (1989). Primate Visions: Gender, Race, and Nature in the World of Modern Science. New York, London: Routledge.

Hulme, M. (2009). Why We Disagree about Climate Change: Understanding Controversy, Inaction and Opportunity, Cambridge University Press, Cambridge.

Jarvis, P. (2012). *Adult learning in the social context.* London: Routledge.

Kolb, D.A. (1984), *Experiential learning: Experience as the source of learning and development*, Englewood Cliffs, N.J.: Prentice-Hall, London.

Leal Filho, W. (2010). Climate Change at Universities: Results of a World Survey. In *Universities and Climate Change – Introducing Climate Change at University Programmes* (pp. 1-19). Berlin: Springer.

Michie, S., Johnston, M., Francis, J., Hardeman, W. & Eccles, M. (2008). From theory to intervention: Mapping theoretically derived behavioral determinants to behaviour change techniques. *Applied Psychology* 57(4), 660-680.

Nahapiet, J., & Ghoshal, S. (1998). Social capital, intellectual capital, and the organizational advantage. *Academy of Management Review* 23(2), 242-266.

Pérez Salgado, F., de Kraker, J., Boon, J., & Klink, M. van der. (2012). Competences for Climate Change Education in a Virtual Mobility Setting. *International Journal of Innovation and Sustainable Development* 6(1), 53-65.

Rogers, E. M. (1995). *Diffusion of innovation*. New York: The Free Press.

Schon, D. (1987). *Educating the reflective practitioner*. San Francisco: Jossey-Bass

Stam, C. D. (2007). Making sense of knowledge productivity: beta testing the KP-enhancer. *Journal of Intellectual Capital* 8(4), 628-640.

UNESCO. (2010). Climate Change Initiative for Sustainable Development. ED-2010/WS/41. Paris: UNESCO.

UNESCO. (2012). *Paris OER Declaration*. World Open Educational Resources Congress. Paris: UNESCO.

United Nations Decade of Education for Sustainable Development (UNDESD, 2005-2014). Retrieved from: *http://www.unesco.org/new/en/education/themes/leading-the-internatio nal-agenda/education-for-sustainable-development/*.

Van der Klink, M. (2012). Professionalisering van het onderwijs. Bekwaam innoveren voor een toekomstbestendig hoger beroepsonderwijs. Inaugural address. Heerlen: Zuyd Hogeschool, university of applied sciences.

Van der Klink, M., Schlusmans, K. & Boon, J. (2007). Designing and implementing views on competencies. In Sicilia, M. *Competencies in organizational e-learning. Concepts and tools*, 221-233. Hersey: Information Science Publishing.

Voorberg, W., Bekker, V. & Tummers, L. (2013). Embarking on the social innovation journey: A systematic review regarding the potential of co-creation with citizens. Paper presented at the IRSPM Conference, Prague, April, 10-12

Willems, P., Kroeze, C., Löhr., A, (2012). The essential role of expertise on natural resources in climate change Master's education. *International Journal of Innovation and Sustainable Development* 6(1), 31-42.

Wilson, G. (2011). "The lived experience of climate change." *Module 2 of the European Union Erasmus project: The lived experience of climate change: e-learning and virtual mobility*. Retrieved from: *http://labspace.open.ac.uk/course/view.php?id=8168*.

Wilson, G., Abbott, D., de Kraker, J., Pérez Salgado, F., Terwisscha van Scheltinga, C., & Willems, P. (2011). The lived experience of climate change: creating open educational resources and virtual mobility for an innovative, integrative and competence-based track at Masters level. *International Journal of Technology Enhanced Learning* 3(2), 111-123.

Wolfers, M.E.G., Van den Hoek, C., Brug, J. & De Zwart, O. (2007). Using intervention mapping to develop a programme to prevent sexually transmittable infections, including HIV, among heterosexual migrant men. *BMC Public Health* 7, 141.

II.
New ICT tools, materials and teachers skills

Let's Play! Using simulation games as a sustainable way to enhance students' motivation and collaboration in Open and Distance Learning

Daniel Otto[1]

Abstract

Simulation games in higher education can be thought of as an innovative tool in order to illustrate complex problems and make them more intelligible. As a student-centered approach it allows personal interaction which enhances student motivation and collaboration. Numerous studies show that emotional and personal commitment can result in better learning outcomes. While simulation games in conventional universities can look back upon a comparatively long tradition, Open and Distance Learning (ODL) for a long time did not take on board this idea. However, recent developments in technology have bridged the gap of adequate tools which have long been identified as the central hurdle in utilizing simulation games for ODL. Based on its characteristic features, simulation games seem particularly suitable for learning about environmental topics. First, it allows students to virtually collaborate and learn about environmental problems in a playful manner which can lead to mutual learning experiences. Second, the concept itself is sustainable as it brings together students without environmental externalities. Empirically, this article draws on experiences gained during an online simulation course of international climate change negotiations. In these negotiations groups of students take the role of states to bargain for a common climate change agreement. Data and student surveys collected since 2010 show that the course led to enhanced motivation and comprehensibility of environmental negotiations processes.

For the institution, it is conceived as a sustainable way to virtually bring together students from different regions and countries. Furthermore, the often lamented isolation in distance learning can be overcome.

Introduction

Simulation games in higher education can be thought of as an innovative tool in order to illustrate complex problems and make them more intelligible. Not surprising, the basic idea to use simulation games for training and education purposes can look back on a comparatively long history. The idea to act out real situations in an easy, comfortable and playful setting originates from military to train future scenarios (Balikci, 2012: 12). To run through scenarios which might occur in reality provides learners with an improved anticipation, leads to enhanced stress control and a better mental focus. This type of simulation games is predominantly used to train professionals for knowledge application.

However, one might also think of using simulation games to better understand complex phenomena and interrelations. This setting applies especially to the context of higher education. Didactically, simulation games are then used as a dynamic and interactive setting which

---

1   FernUniversität in Hagen, 58084 Hagen, Germany.

stimulates collaborative learning. On the one hand in higher education, students regularly learn about complex problems that require a great capacity for abstraction. On the other hand, students often tend to underestimate the difficulty behind ostensibly informal processes. In both cases, simulation games can be seen as an adequate tool to present the complexity of a given problem in an innovative way, different from the regular curriculum. Furthermore, simulations games are not limited to solely knowledge transfer but deliver fruitful insights by playfully opening up new perspectives.

Consequently, this paper is based on the assumption that simulations games offer numerous benefits for students by learning in interactive collaboration. This assumption is supported by the literature about simulations games (Breitmeier and Otto, 2012; Powers and Kirkpatrick, 2013). The replication of a real situation in a given setting helps to better understand settings of different levels of complexity. To apply knowledge in these situations stimulates the student's learning motivation and learning outcomes.

A glance at the literature on simulation games reveals that there is no clear use of the terms simulation and gaming (Crookall, 2010). Often, simulation games are thought of as computer games which are played for various purposes, e.g. war games and role plays. Ambitious graphics and sophisticated gaming experiences are consequently necessary features. It is important to point out that simulation games are not necessarily equal to computer games. Quite contrary, simulation games are ever so often realized without technical support in face to face settings (Breitmeier and Otto, 2012: 28). In their literature review Lean et al. underline the broad heterogeneousness of the existing approaches (Lean et al., 2006: 228). However, the simulation games we refer to in this paper are computer based but not in an ambitious manner as they do not require complex computer programming services or disproportionate technical efforts.

This definition puts emphasis on the potential of simulation games to reflect reality and thus includes all games which support this fundamental idea. For that reason games are understood as an electronic learning setting accessible via electronic devices. Contrary to games, we distinguish simulations as a method to run through a given setting which is inspired by reality. Although reality is often too complex to be reproduced exactly, it is the initial point for starting a simulation game. In the literature the focus is often on the technical aspects of simulation games by neglecting its intent and purpose that is to stimulate learning through interaction and collaboration.

At first sight simulation games seem to offer several benefits for higher education. In practice, teachers and tutors are often sceptical about integrating simulation games into their curriculum. Compared to its potential benefits, there is little use of simulation games in higher education (Kovalik and Kuo, 2012). Initiatives to adapt simulations for teaching and learning are often seen as contradictive to the classical didactical model. This model is based on the premise to initiate an information transfer directly from educator to recipient (Ruben, 1999). Simulation games on the contrary follow a student-centered approach to stimulate personal interaction and to enhance the student's motivation and collaboration. Numerous studies show that this emotional and personal engagement can result in better learning outcomes (de Freitas, 2006).

Against this background, simulation games have entered the discussion about refreshing the university's curricula. Studies revealed that teachers and tutors who have used simulation games think of them as valuable for their learning and teaching practices (de Freitas, 2006). While simulation games were able to find its way in conventional universities, Open and Distance Learning (ODL) for a long time did not take on board this idea. One central explanation for this reluctance might be the lack of adequate hard- and software to realize simulation games in a distance learning setting.

Recent developments in technology triggered by Web 2.0 have bridged the gap of adequate tools and therefore taken the central hurdle to utilize simulation games in ODL. New ICT tools like enriched virtual learning environments (VLE), videoconferencing, wikis, podcasts etc. offer various forms for new modes of communication.

To sum up, the before mentioned simulations games can be advantageous for both students and universities (table 1):

Table 1:   Advantages from simulations games for students and universities

Students have benefits because simulation games:
1. *Offer a collaborative effort.* Students often lament the isolation associated with distance learning. Collaboratively organized simulation games can overcome this feeling by creating an online group for active learning. Furthermore, research shows that simulations can be deployed as icebreakers and team-building exercises (Anderson and Lawton, 2008).
2. *Enhance motivation.* Interactive group-learning situations instead of solely studying written materials can increase student's motivation to complete a course.
3. *Develop soft skills.* Especially group oriented simulation games require a high amount of communication and team-orientated working from the participants. To put oneself for example in a country's position, group discussions and to bargain for agreement will require empathy, flexibility and agreement.
4. *Enhance understanding of real world situations.* Although simulation games can never duplicate real world situations in its complexity, it draws upon these situations to better understand its relevance and difficulty. To learn about processes which otherwise can only be observed from the outside in best case leads to a reassessment of thought processes.
Universities providing simulation games will benefit because simulation games:
1. *Refresh and extend the curricula.* Simulation games are a feasible way to modernize curricula with attractive alternatives to regular courses. According to this, they can be a recognition value for students.
2. *Cost friendly.* Simulations as I define them in this paper can be established at comparatively low use of resources as they do not require complex programming. If one refrains from fastidiously computer games, a simulation game can be developed with usual ODL software.
3. *Bridge the distance.* Although ODL is designed as distance teaching, attendance seminars are often used complementary at the beginning or end of a course. Students from abroad regularly have problems to attend these seminars. Simulation games put students in the position to virtually come together at low costs.
4. *Are environmental sustainable.* As mentioned above the idea of simulation games in ODL is a sustainable concept, as it brings together students without producing environmental externalities.

All in all, it can be stated that simulation games are a fruitful and sustainable way for ODL and offer several benefits for students as well as universities which provide them.

In the following we want to make a jump from theory to practice. Still too many discussions about simulation games in ODL are stuck in theoretical reflections. Bearing on personal experiences I would like to revive the practical side and take a stance for the active application of simulation games in ODL. The course we present is called *international climate change negotiations seminar (iccn).* iccn is not presented as the "answer to everything" but one of a thousand ways to translate simulation games into ODL. It serves as an example of how simulation games can lead to tangible impacts at appropriate costs. Further, it identifies obstacles and indications for realizing simulation games in one's own curriculum. iccn is offered in our Master program called *Interdisciplinary Environmental Science.* Target groups are employees with a first degree and a working expertise in any environmental sphere. The curriculum is fully at distance, only providing weekend attendance seminars at the end most courses. While the curriculum is mainly based on classical course reading material, it also offers an online course. The online course is designed for two groups of students. Firstly, students who live abroad and thus are not able to be present at attendance seminars. Secondly, students who are interested in group-oriented learning.

Based on its characteristic features, simulation games seem particularly suitable to learn about environmental topics. Climate change is the superior environmental problem of the recent decades as its impacts affect all kind of spheres of life. For understanding climate change, interdisciplinarity is required to fully reflect all its dimensions. Even though natural science informs about the causes and impacts of climate change, politics is responsible to secure effective problem solving and responsiveness (Breitmeier and Otto 2012). On the international level, since 1995 climate change negotiations are organized as annual meetings where state actors bargain for agreement. In 1997, this led to the Kyoto-protocol where for the first time binding reduction targets were codified. Although most of the students are aware of these negotiations via the media, they often lack understanding for its poor outcomes. Students regularly complain about the inability of political representatives to bargain compromise and to adequately deal with the problem.

Enabling students to switch the sides to better understand the dynamics of international climate change negotiations is the basic idea of iccn. As a side effect, students also acquire in-depth knowledge about the politics of climate change. Climate change negotiations take place every year in a two week period accompanied by several meetings throughout the year which encompass diverse sub-topics. This complexity cannot be fully reflected in a four months distance teaching course. Therefore the course is dedicated to a special aspect of the climate change negotiations. In the last years REDD (Reducing Emissions from Deforestation and Degradation) has been the topic of choice. Avoided deforestation and degradation in developing countries and its benefits has received incremental prominence in the recent negotiation rounds. During iccn students play an international conference on REDD with the aim to negotiate a common agreement suitable for all participants. To allow feasibleness the number of participants is limited to 25 students.

Structure of the course

The course is divided into three main parts:

1. *Preparation phase.* During preparation phase, students receive an introduction to the course and get access to the virtual learning platform (Moodle based). For participation, students first have to write a 5 to 10 pages research paper. One half of the paper is about REDD in general; the second is about the negotiation position of one specific country towards REDD. This procedure secures that all participants have an idea about the concept of REDD and why it is a problem in the country under investigation. Helpful literature is provided in the platform. Special weight is put on literature about negotiations and its strategies. All research papers are graded to give a first feedback on the performance level. Afterwards the papers are made accessible to all member of the course. Beyond, students are informed about the structure of the course, the technical requirements and can ask questions to the tutors.

2. *Composition phase.* Only students who have successfully finished their research papers will be admitted to the second stage. In the last three years the drop-out rate to the second stage was very low with only two out of 75 students. The composition of the country

groups is the task of the second stage. Granted that there is a maximum participation of 25 students, they are divided into five groups. Each group takes the position of one country. Usually, the United States and Europe are included as influential members. They are complemented by one transition country like for example Brazil and two states which will be heavily affected by changes in the REDD policy, for example Indonesia and Ghana.

3. *Negotiation phase.* After the groups are composed the actual negotiations start. The negotiations are structured into six sessions, one each week at the same time. Technically, the sessions are done with Adobe Connect software. Adobe Connect allows for several features like creating a virtual meeting room, to separate students in subgroups, speech-based communication, the use of whiteboards etc.

In the *first session*, all participants meet in the plenum. All sessions are supervised by me and another tutor. After an introduction the students are divided into their country groups to prepare a common negotiation position. This position is presented in the second session in the form of a written statement. All statements have to be posted in the virtual platform two days before the next session. This procedure facilitates substantial discussions. Communication in the sessions is done via headset while it is optional to use a webcam. During the group work, we virtually jump from group to group to follow the discussion and intervene when needed. Groups can also call us if they feel that assistance is needed. Before the session is over, all participants meet in the plenum to clarify open questions and receive instructions for the next session.

In the *second session*, all groups meet in the plenum to present their country's position on REDD. Like in real negotiations, each group is given the same time for its presentation. Afterwards students are divided into their groups and have time to prepare comments on the positions of the other groups. This is again done in the plenum with strict time management.

For the *third session*, two ad-hoc groups are created which each discussing one specific aspect for a REDD agreement. Each country group nominates representatives for every ad-hoc group. After a first consultation in the country groups to agree on a common negotiation strategy, they send representatives to the ad-hoc groups to bargain compromise. The ad-hoc groups all vote a moderator and secretary who records the outcomes. The secretaries post the outcomes in the learning platform during the next week.

In the *fourth session*, the country groups debate the outcomes of the ad-hoc groups. They now have the chance to reformulate, adapt or expand their position. In the plenum, the results of that discussion have to be spread to the other country groups. The final written comments have been uploaded in the learning platform by then.

The *fifth session* is again a meeting of the ad-hoc groups. After a short meeting in the country groups the ad-hoc groups are established to continue negotiations.

Before the *sixth session*, the skills of the tutors are required. We bundle the discussions of the ad-hoc groups and produce a proposal for referendum. This proposal is uploaded in the virtual learning platform before the sixth session. During the session each country group votes whether it agrees or disagrees to the elements in the proposal and if they can accept it. An elected speaker votes and comments for each country group. We take the role of chief negotiators. Typically, while country groups agree on some points, it is impossible to find full consensus on the proposal. The session is closed with some concluding words by us. We summarize the main outcomes and give a short reflection about negotiations and its dynamics.

4. *Individual and group-reflection-phase.* The negotiation phase is followed by a reflection-phase in which students individually and collectively reflect on their experiences and les-

sons learned during the course. With this notion we refer to Petranek and his concept of written debriefing in the context of simulation games which is understood "as an experiential activity in which participants have the opportunity to write about their experiences and feelings and those of others" (Petranek, 2000: 109). Petranek claims that the outcome of writing down experiences is superior to oral debriefing as it permits to initiate a process of in-depth self-reflection and in the end learning. The individual part is a two page paper where students write down their individual experiences and state a compliment or a criticism for the course. For the group-reflection a forum is released in the virtual learning platform. Students have the chance to discuss the outcomes of the negotiations or the course in general.

Table 2:   Structure of the course

	Task	Duration	Participants
Preparation phase.	Produce research paper	One month	All students
Composition phase.	select group	One week	Students who passed stage one
Negotiation phase.	Start of the simulation	Two month/six sessions	Students who passed stage one
Individual and group-reflection-phase.	Write report and discuss in learning platform	One month	Students who passed stage one

Evaluation

For us as the heads of the course it is important to receive feedback on the student's opinion. This is especially the case because the course was established on our own thoughts against the background of academic and public literature on simulations and games. We do not understand the course as fixed but rather as a dynamic process based on an action-reflection-learning-action-cycle. The individual and group-reflection-phase therefore is not for the sake of it, but to receive guidance to further develop the course. Methodically, we prefer qualitative instead of quantitative evaluation. We are convinced that qualitative evaluation based on written debriefing allows for a more in depth analysis. To write freely gives students the chance to self-reflect and to cultivate their own thoughts based on gained experiences. Unaffected by the subtle implications of questions, critical thinking and setting own focusses is possible. However, the guidance to the reflection-reports holds some food for thought of what could be included in the report.

The course can be attended since 2011. Integrated in the Master's regular curriculum, it is offered once a year. As mentioned before, the number of participants is limited to 25 for each course. Due to the fact that the individual and group-reflection-report is obligatory for completing the course, we received a participation rate of approximately 100 percent. Thus, the following evaluation reflects three courses with a total of 73 participants. Although we made small course modifications within that period, the basic framework of the course remained constant. Our evaluation of the 73 group-reflection-reports is based on content analysis methodology. Statements and opinions about the course in the reports were classified based on inductively derived negative and positive categories. In order to ensure intercoder reliability, the content analysis was conducted independently by each course leader to unveil the amount of agreement and thus provide a solid test.

Table 2: Numbers from the the course

Course year	Participants	Students who completed	Number of drop-outs	Success rate in percentage
2011	25	24	1	96
2012	25	24	1	96
2013	25	25	0	100
total	75	73	2	97.3

In the following, we want to highlight the most prominent distinctive features, regularly expressed by the students. I will consider the individual-reflection-reports as well as the group discussions in the forum. With our comments we hope to contribute to the further development of simulation games in the context of ODL and encourage educators to make use of its benefits. At the same time I want to point out hurdles which have to be taken into consideration.

Awareness-raising for REDD and climate change

Nearly all students agreed that the iccn has increased their sensibility for the problem of climate change and the international negotiation process on REDD. Although all students had basic knowledge of environmental topics, almost everybody of them had never heard of REDD. Students expressed that the course has raised their interest in REDD and that they would now be an attentive observer of the ongoing process.

These statements show that the course was able to raise awareness for the topic of climate change negotiations. As many students indicated to follow up on the further process this awareness seems to be sustainable.

Role play as an active and realistic approach to negotiations

Our idea to use role play as the didactic core was appreciated by the students. They fancied the idea to take the positions of states and bargain for agreement. Interestingly, most students found it difficult to overcome their "Germanized" position. Laggard states like the US contradicted their normative position as environmentalists. This was also evident in the choice of the country groups where the US was continuously the last country which was selected. Nevertheless students appreciated the challenge to question their normative positions by putting oneself in the position of another country. The opportunity to act on behalf of real state positions was seen as exiting.

Secondly, it was surprising for us to discover that students felt they were given realistic insight of the dynamics of international negotiations. Even though there were three dissenting voices, the overwhelming number of students shared this impression. Catchwords often named were time pressure, vital discussion and consensus finding, deadlocks, pending between groups and compromise. As political scientists we are aware that negotiations can never be fully reflected in a three month course. However, we share the impression that the dynamics of this process can be illustrated at its core. This tendency is increased when students lack previous knowledge about negotiations and REDD.

Remove isolation encourage collaboration

An issue frequently highlighted in the individual reflection report was the collaborative character of the course. Interaction and the exchange with others, group-working and video-
conferencing was often named as benefits. As one student put it, this course was in total
contradiction to the classical idea of submitting written assignments. Due to successful group
collaboration, some students planned to initiate regular virtual or physical exchange.

Generally speaking, in distance learning students often find themselves isolated and lack
the interchange with like-minded students. Courses which include group-working and video-
conferencing encourage students to break this isolation. This does not only relate to professional but also to private exchange.

Virtually crossing borders with maximum flexibility

Distance education means to have students with various backgrounds. In a master program
which is based on the idea of interdisciplinary and where working experience is required, this
variety is intensified. Our students mostly work in the environmental sector often located in
developing and transition countries. This target group, not able to participate in attendance
seminars, made up the biggest number of participants in the course. This group appreciated
the chance to take the course while full-time working abroad without having externalities. In
principle, all students benefited from this course structure, but empirically, this benefit was
strikingly mentioned by students working abroad.

After listing four major benefits, in the following we focus on negative comments of the
students and discuss two salient points of criticism. These two are highlighted because we
agree that they are central hurdles which have to be taken into consideration for simulations
games in ODL.

Expenditure of time

Conducting this type of course is time consuming for both students and teachers. Teachers
need a lot of time to prepare and to be present during the sessions. In addition, research papers
have to be graded in advance to the online sessions. Students, contrary to classical distance
courses, have fixed weekly appointments where they meet virtually. Between the sessions
there must be time devoted to coordinate the groups and to prepare the sessions. The four
month period include writing a research paper, attending, preparing and following on the sessions and writing the reflection-report. Against this background it is no surprise that the course
was perceived as intensive, and as to require a high amount of flexibility and the willingness to
play an active part. Students stated that significant time was needed to prepare and follow up the
sessions, e.g. to write protocols or produce statements. Therefore many groups were meeting
outside the regular sessions using Skype or Email for collaboration. Although students indicated the expenditure of time they all regarded the course as being valuable. For the teacher
an extra time has to be budgeted compared to classical courses.

Technical problems

Whenever there is technic involved it normally causes problems. This was likewise the case
in all three years since the course is running. The main cause for problems is the technical

infrastructure used for the online sessions. While chat based communication is not a problem, speech and video based communication certainly is. The situation is further aggravated because many students do not configure their hard- and software correctly. Beyond that, students were regularly kicked out by the program and had to reconnect. One reason for that is that students were linked in from developing countries where the connection is often not stable. Adding all this up, too much time had to be devoted to fixing technical problems which was lamented by the students in their reflection reports. Although we share this expression of criticism our opinion is that there are no serious alternatives. While it might be an option to change software this will only change the problem but not remove it permanently. Continuous technical development and improvement of soft- and hardware products seems the only solution.

Simulation games in ODL – a theoretical and practical conclusion

Simulations games are a fruitful contribution to ODL. Apart from the classical model of teaching it provides a collaborative approach to learning about environmental topics. Furthermore, it not only enhances the attractiveness of ODL but also leads to tangible learning outcomes. With this article, we want to give empirical evidence to this argument.

While the theoretical argument seems obvious there is a lack of practical experience in ODL. This is surprising because ODL, at least in the recent past, provides all the technical requirements to conduct these kinds of courses. In conventional teaching surveys give evidence that cooperative learning (Herrmann, 2013) and simulation games (Moizer et al., 2009) lead to better learning outcomes. Although our experience supports the assumption that this is also true for ODL, broader empirical evidence is still missing (Bidarra et al., 2011). However, the main hurdle seems to be a misunderstanding of simulation games. Far too often simulation games are thought of as complex and technically sophisticated products. This may take the biscuit for several contexts but is not necessarily the case for ODL. If the basic rationale of simulation games is understood by means of representing a subject-matter of reality in a virtual surrounding it does not require these features. Derived from our three years of experience I firmly believe that the interactive and teamwork character is far more appreciated than a pretty shell, therefore, "keep it simple and play!"

Acknowledgements

I thank Katharina Frank and Alexander Bollmann for their helpful comments.

References

Anderson PH and Lawton L (2008) Business Simulations and Cognitive Learning: Developments, Desires, and Future Directions. *Simulation & Gaming*, 40(2), 193-216, Available from: *http://sag.sagepub.com/cgi/content/long/40/2/193* (accessed 5 August 2013).

Balikci A (2012) Das systemische Planspiel. Lernen durch Erleben. *Sozial extra* 8/10, 12-14.

Bidarra J, Rothschild M and Squire K (2011) *Computer Games as Educational and Management Tools.* Cruz-Cunha MM, Varvalho VH, and Tavares P (eds), IGI Global, Available from: *http://www.igi-global.com/chapter/games-simulations-distance-learning/53951/* (accessed 12 February 2014).

Breitmeier H and Otto D (2012) Understanding Political Processes in Climate Change Negotiations by means of an Interdisciplinary Curriculum in Higher Education. *International Journal on Innovation and Sustainable Development* 6(1), 20-30.

Crookall D (2010) Serious Games, Debriefing, and Simulation/Gaming as a Discipline. *Simulation & Gaming*, 41 (6), 898-920, Available from: *http://sag.sagepub.com/content/41/6/898.abstract*.

De Freitas SI (2006) Using games and simulations for supporting learning. *Learning, Media and Technology*, Routledge, 31(4), 343-358, Available from: *http://dx.doi.org/10.1080/17439880601021967*.

Herrmann KJ (2013) The impact of cooperative learning on student engagement: Results from an intervention. *Active Learning in Higher Education* 14 (3), 175-187, Available from: *http://alh.sagepub.com/content/14/3/175.abstract*.

Kovalik CL and Kuo C-L (2012) Innovation Diffusion: Learner benefits and instructor insights with the DIFFUSION SIMULATION GAME. *Simulation & Gaming* 43(6), 803-824.

Lean J, Moizer J, Towler M, et al. (2006) Simulations and games: Use and barriers in higher education. *Active Learning in Higher Education* 7 (3), 227-242, Available from: *http://alh.sagepub.com/content/7/3/227.abstract*.

Moizer J, Lean J, Towler M, et al. (2009) Simulations and games: Overcoming the barriers to their use in higher education. *Active Learning in Higher Education* 10 (3), 207-224 Available from: *http://alh.sagepub.com/content/10/3/207.abstract*.

Petranek CF (2000) Written Debriefing: The Next Vital Step in Learning with Simulations. *Simulation & Gaming* 31(1), 108-118 Available from: *http://sag.sagepub.com/cgi/content/long/31/1/108* (accessed 5 August 2013).

Powers RB and Kirkpatrick K (2013) Playing With Conflict: Teaching Conflict Resolution Through Simulations and Games. *Simulation & Gaming*, 44 (1), 51-72, Available from: *http://sag.sagepub.com/content/44/1/51.abstract*.

Ruben BD (1999) Simulations, Games, and Experience-Based Learning: The Quest for a New Paradigm for Teaching and Learning. *Simulation & Gaming*, 30(4), 498-505, Available from: *http://sag.sagepub.com/content/30/4/498.abstract*.

Developing E-Learning Materials for Teaching Industrial Ecology and Environmental Sustainability

Anthony Halog and Gary Dishman[1]

Abstract

In support of the urgent call for climate change mitigation and pursuing the vision of sustainable, circular economy, increasing number of universities have been developing courses to prepare students in meeting the growing demand for green and sustainability related jobs, which require systems perspective training in Industrial Ecology (IE) and Life Cycle Assessment (LCA). LCA analyses emissions and resource extractions from mining and processing of resources (cradle) to disposal of wastes and residuals (grave) and even back to cradle. This paper plans to contribute in developing an online-based information system which provides an interdisciplinary approach to teach IE and LCA for sustainable industrial development. Specifically, the aims of this e-learning project are: (1) To develop innovative teaching and learning materials in line with flipped classroom strategies and (2) To create a one-stop Blackboard in support of learning the principles and concepts of IE and LCA for Sustainable Production and Consumption. Though the developed materials are used first to support face-to-face learning, this is being leveraged to meet demand for distance/online LCA education in the near future.

Introduction

The effective implementation of sustainable development strategies is essential in order to address the environmental challenges which threaten our future (UN DESA, 2008). Challenges such as climate change are not only environmental or sustainable development issues, but global social concerns (UN DESA, 2008). Education can be seen as a tool to address this societal need for better understanding of sustainable development practices and related environmental concepts (Barth and Burandt, 2013). In addition, effective education strategies can be utilised to allow students to develop a broad range of sustainable competencies that can be applied in their future careers (Bielefeldt, 2013).

In response to the need for a society who understands sustainable development practices, there has been an increase in the demand for professionals who have developed environmental expertise in areas such as Industrial Ecology (IE) and Life Cycle Assessment (LCA). Universities, such as the University of Queensland (UQ), have adapted to change by developing programmes that educate students in these environmental competencies. Within UQ, the School of Geography, Planning an Environmental Management (GPEM) has a number of programmes designed to create the next generation of environmental decision makers and scientists. The mission of the school is:

---

1   School of Geography, Planning & Environmental Management, University of Queensland, St Lucia, Australia 4072.

"To understand the processes, structures, and interfaces of natural and human systems and to use this knowledge to create and inform the next generation of knowledge leaders and to help shape policy, planning and management to enhance sustainability". (GPEM, 2013)

With this in mind, a number of their programmes have been specifically designed to include courses that can train students to understand the technical competencies of LCA and IE whilst obtaining the analytical skills to successfully contribute to sustainable industrial development. Within the school, the two courses that focus on IE and LCA are: Sustainable Consumption and Production (ENVM2101) and Industrial Ecology and Life Cycle Thinking (ENVM3528). ENVM2101 focuses on theories and strategies around the sustainable consumption and production of goods and services. ENVM3528 builds on the core learning of ENVM2101 and takes a life cycle thinking approach with an aim to develop enhanced technical competency in Life Cycle Assessment (detailed course descriptions can be found in the 'Background to the Project' section).

In addition to the traditional pedagogical methods of physical lesson delivery, there has been recent interest in how e-learning technologies can be utilised in teaching sustainability concepts (Barth and Burandt, 2013) and there is much potential in e-learning to offer innovations in education. Barth and Burandt (2013) do note however that utilising e-learning in delivering sustainability education does not come without challenges. It is therefore necessary to take a step-by-step approach in applying e-learning methodologies and aim to gradually implement new innovations to help better understand their performance.

This chapter follows this approach and presents a constructive analysis of how e-learning methodologies were applied to the teaching of sustainable industrial development. From applying the concepts we can derive best practices that can be utilised in the future delivery of online/distance learning LCA education. The chapter briefly looks at how education can be used to promote sustainable development as well as relevant e-learning methodologies before analysing how the concepts were applied to the ENVM2101 and ENVM3528 courses. Following on from application is a presentation of outcomes and recommendations that can guide pedagogical methods in the delivery of sustainable industrial development education.

The Need for Effective Sustainability Education

Education has always been a key element in developing strategic approaches and policies guiding sustainable development (Barth and Burandt, 2013). In relation to this, the idea of education for sustainability (EfS) emerged with the view to support quality education which is inclusive of all individuals that can provide the knowledge required to address current and future sustainability challenges (UNESCO, 2009; Barth and Burandt, 2013).

Sustainability education has become even more prominent within higher education over the past few years with a focus on encouraging ongoing awareness and understanding of environmental concepts (Stewart, 2010). As mentioned previously, a deeper understanding achieved through effective education could enhance society's appreciation of sustainable development concepts and potentially future actions. It was realised that courses could act as 'agents of change' in developing a sustainability mindset that students could apply in their everyday lives and future careers, whilst impacting a wider audience (NCSE, 2003; Stewart, 2010; McNamara, 2011). Taking this into account, one can consider that the provision of sustainability education as both a tool for delivering the theoretical concepts and something that can elicit future behaviours that focus on critical analysis and application. Barth and Burandt (2013, p. 5) identify the capacity of Education for Sustainability (EfS) to allow individuals to: *"not only to acquire and generate knowledge, but also to reflect on further issues such as the complexity of behaviour and decisions in a future-oriented, global perspective of responsibility.".*

The authors identify this as a paradigm change and support this view with literature focusing on; sustainability education being *about, in* and *for* sustainability (Sterling, 2001) and the idea that in one dimension EfS supports particular behaviours or methods of thinking and in another will encourage critical analysis and a mindset that wishes to explore contradictions in a sustainable environment (Vare and Scott, 2007). In taking this approach we have acknowledged the fact that our courses must deliver the material that will both educate and challenge students by giving them the skills to think critically about a diverse range of concepts. In a sense what we are trying to achieve goes beyond developing understanding, but creating a mindset that can fully appreciate the complexities in the sustainable development realm. The behaviours that we are trying to elicit will also be complemented by the enhancement of competencies that can support them well beyond the initial delivery of the course. One could argue that what we aim to give students is more than just a lesson, but an 'experience' which is fully supported by effective course design, clearly defined objectives, competency development and the promotion of a sustainable development mentality.

The realisation that sustainability education requires much more than just teaching the concepts lends itself to investigating how other methods can support a holistic understanding and appreciation of sustainable development. Indeed, Barth and Burandt (2013) acknowledge this by noting the fact that there are a diverse range of teaching methodologies and best practices that create effective learning environments including experiential-learning (Brundiers et al., 2010) and problem-based learning (Brundiers and Wiek, 2013; Thomas, 2009). However, much of the work performed by Barth and Burandt in their 2013 paper explored the benefits and challenges of applying e-learning concepts to a number of sustainable development case studies. From our point of view such work offers numerous opportunities and helps support the need to further understand how we can create an effective learning experience for sustainable development by utilising e-learning technologies.

Understanding E-Learning Concepts

The internet, along with information and communication technologies, are becoming an integral part of our lives on a global scale as we progress within the information age (Katz and Rice, 2002; Mossberger, Tolbert and McNeal, 2008; Castells, 2011). The advent of the internet has increased the application of technological methodologies on education through facilitating improved collaboration and the exchange of resources (Walton et al., 2008). Although the internet and communication technologies have been applied with great success to many different areas, it can be argued that e-learning in teaching sustainability may require a newer dynamic approach, with much left still to be explored (Bielefeldt, 2013). The case study aims to add to the body of knowledge regarding sustainability education, but before that we need to outline some of the e-learning concepts that offer potential for further investigation and application. The three concepts of most interest are open learning environments, flipped learning and the recent developments in Massive Open Online Courses.

Open Learning Environments

Open learning environments are a learner centric model, which is based on authentic learning situations, providing context, a starting point and framework for the whole learning setting

(Barth and Burandt, 2013). Barth and Burandt (2013) suggest that, through open learning environments, control is given to the learner through processes that allow for self-guidance and tools that facilitate individual and collaborative exploration of concepts. Following on from this, Barth and Burandt (2013) explain the design of such learning environments, which can be embodied by three fundamental principles:

- **Self-directed learning**. This focuses on the individual learner through which they acquire knowledge and skills in a manner that is self-managed. Individuals learn, not through a taught transfer of concepts, but via ongoing autonomous knowledge discovery;
- **Collaborative learning**. Collaboration is a social activity that focuses on engagement between different actors. These actors, which may include the students and teachers, have the same learning objectives which are achieved through joint participation and experience;
- **Problem Orientated Learning** focuses on the development of competencies that derive from solving real-life challenges and scenarios. Learners have to create effective solutions to such challenges and in turn they acquire an understanding of how they may react in related situations.

It can be argued that the philosophy of open learning environments is perfect for creating an 'experience' by which the students are in control and work with others to enhance their understanding of sustainable development concepts. eLearning lends itself perfectly to open learning environments as it incorporates tools and functionality to support engagement among stakeholders and the delivery of content to support individually focused learning (Barth and Burandt, 2013).

Flipped Learning

Leading on from open learning environments is the relatively new philosophy of flipped learning. Flipped learning is a concept whereby the focus shifts from teacher centric, in class methods, to an ideology which focuses around student-centric, self-directed learning (Tucker, 2012). This concept is in essence 'flipping' the responsibilities (Strayer, 2007). The core idea identified by Tucker (2012, p. 82) is as follows:

> *"With teacher-created videos and interactive lessons, instruction that used to occur in class is now accessed at home, in advance of class. Class becomes the place to work through problems, advance concepts, and engage in collaborative learning. Most importantly, all aspects of instruction can be rethought to best maximize the scarcest learning resource – time."*

A typical application would be to utilise the vast video resources available online to create a repository of learning materials for students which they can review in their own time (Knewton, 2012; Szafir and Mutlu, 2013).A flipped learning approach also allows the students to test their theoretical understanding and utilise tools that facilitate peer engagement and knowledge application (Knewton, 2012).

Flipped learning from the point of enhancing a sustainability mindset and delivering a true learning experience offers much potential. A core focus is on engaging the student from home before they have attended the class through utilising instructor-led videos and other related content and activities. This in itself perfectly matches the principles of open learning environments and supports the ongoing acquisition of knowledge and competencies from beyond the realm of the classroom. Another aspect that flipped classrooms supports is collaborative learning through physical in-class contact and through online social tools. It can be argued that by utilising e-learning tools and creating new opportunities for interaction amongst learners and tutors, students may be able to enhance their understanding of sustainable development concepts and be able to develop critical analysis skills in a collaborative environment (Barth and Burandt, 2013). Another concept of flipped learning is that of supporting a problem orientated approach through the delivery of online exercise, assignments and activities. Tucker (2012) noted that those who had implemented flipped learning strategies saw an increased motivation to work through challenging problems and think more deeply about the concepts. In terms of course development, the ability to facilitate exercises to support problem orientated learning and an increase students applying themselves to these challenges, may help in creating enhanced action competence in sustainable development scenarios. Some authors (i.e. Janz et al., 2012; Tucker, 2012) however have noted that there are challenges in implementing flipped learning strategies and a level of understanding is required in to effectively integrate traditional pedagogies with new teaching concepts. From evidence gathered from those who have implemented flipped learning, Tucker (2012) acknowledged the fact that developing short instructional videos to accurately represent the learning objectives is a significant challenge in course design. Janz et al. (2012) noted that a level of stakeholder buy-in and collaboration from learners and instructors is required to effectively implement flipped learning. Despite these challenges, flipped learning offers much potential for enhanced student engagement in a multi-channel delivery of sustainability focused education

Massive Open Online Courses (MOOCS)

One additional consideration which is of future concern is the recent development of Massive Open Online Courses or MOOCs. There has been much focus on MOOCS and is a buzzword that has recently entered the educational mainstream (Daniel, 2012). In essence a MOOC is an online course that allows for massive participation and online engagement. With regard to the paper, the potential to deliver sustainable industrial development theory online is a potential evolution from what we are currently trying to achieve. In designing the learning environment for the courses we will consider MOOCs with a view to incorporate videos and online content in future developments. As we have highlighted, sustainability education is both growing in importance and evolving and anything that can be applied to support the delivery of a sustainability mindset must be appreciated. It must be acknowledged however that MOOCs are a relatively new concept in world of online education and its overall effectiveness for developing knowledge is still being examined (Daniel, 2012). Daniel (2012) also noted that there is no universal standard in developing MOOCs and that aspects such as quality, student attrition, financial justification and certification may need to be explored. In the future, any implementation will look to acknowledge these questions and apply findings from current course development in order to enhance the performance of a sustainability MOOC further down the line.

Background to the Project

Having understood the fundamentals behind teaching sustainability education and the e-learning tools can that facilitate new and innovative ways of delivering material, we can now apply these concepts to designing our approach to the courses. In order to better understand what we are trying to achieve we must first analyse the courses themselves.

The two courses both aim to deliver an understanding of core environmental concepts with a focus on sustainable activities and life cycle assessment. Both of the courses follow a format of 1 two hour lecture in conjunction with 1 one hour tutorial over the duration of 13 weeks. The

Sustainable Consumption and Production Course is designed as core learning for environmental programmes offered at UQ and to provide an underlying base for students wishing to develop their expertise in a particular environmental field. A description of the course is as follows:

> "*The production and consumption of goods and services is central to many of the sustainability problems currently faced by our society. This course introduces students to consumption patterns and production processes using a life cycle perspective and examines why they can be unsustainable. It then examines the various strategies employed to bring about more sustainable production and consumption; on the supply-side and the demand-side. These include Life Cycle Assessment, Cleaner Production, renewable energy systems, and communication of environmental information. Emphasis is placed on providing examples of practical applications for a range of production systems- manufacturing, food supply chains, energy generation, etc. Students will gain skills in identifying and critically evaluating solutions for real-life sustainability problems.*"

Industrial Ecology and Life Cycle Thinking is offered as an elective for those who want to specialise and develop further understanding of concepts such as green economy, industrial economy, industrial symbiosis, socio-economic metabolism, circular economy and methods and tools of life cycle and systems thinking. A description for this course is as follows:

> "*The production and consumption of goods and services is central to many of the sustainability related problems currently faced by human society. This course introduces students to the theories and applications of industrial ecology and environmental life cycle assessment (LCA) in natural sciences, engineering, business economic and government policy decision situations. Students will do problem sets, learn how to use LCA software models & tools, review cases and conduct an LCA project such as environmental life cycle assessment of industrial products and systems (e.g. food systems, mining and minerals, renewable energy technologies).*"

The previous iterations of the Sustainable Consumption and Production course adopted a more traditional learning approach that focused on instructor-led lectures and tutorials as well as knowledge resources in the form of journal articles and other text based content. The course used Blackboard to some extent but tools such as forums and embedded videos were not being applied to their full potential. However, previous content and best practice was used as a template to help support initial course design of the current e-learning implementation. The Industrial Ecology and Life Cycle Thinking course is a new offering at the University meaning there is a clean slate for the development and design of learning approaches and Blackboard. The general structure and content for this course had been submitted previously for pre-approval; however the application of e-learning methodologies with focus on sustainability is something that could be developed as the course progresses.

This chapter was written towards the end of 2013 after the initial pilot implementation of an e-learning approach for Sustainable Consumption and Production (ENVM2101) had been completed. Industrial Ecology and Life Cycle Thinking (ENVM3528) is scheduled to run in February to May, 2014. Findings from the initial implementation of ENVM2101 will be applied throughout the future delivery of the ENVM3528 course.

Blackboard

In the development of the ENVM2101 course in the second half of 2013, specific analysis and training in the available tools was conducted. Through previous understanding and new insights gathered, it was concluded that Blackboard provided all the tools required to support a holistic flipped learning strategy and effectively engage with the student audience. This included the functionality to upload and link to content, develop assessment items, post discussion questions and facilitate engagement between both the instructor and other students.

Objectives

From analysing the course profile and understanding potential e-learning approaches, a number of specific objectives were identified. These all fall under the higher level goal of an effective teaching environment to support the theoretical understanding and practical application of sustainability concepts. These objectives particularly focused on developing the skills required to meet the growing demand of LCA trained professionals and providing them with a broad spectrum of sustainable development competencies.

1. A well-developed, high quality Blackboard with relevant web links and access to You-Tube's or personally recorded lecture videos, towards a one-stop online/distance-learning tool. The online course materials include course documents, videos and readings, case studies, direct access to public available inventory data sources, homework and test exercises;
2. Multiple-choice exam questions created within Blackboard. This can support students who are preparing for LCA Certification Exam and interested to be certified by the American Centre for Life Cycle Assessment;
3. Direct and easy access to downloadable LCA software packages (e.g. Open LCA or CMLCA) for applying LCA principles and concepts in student's semester projects.

Benefits

From the objectives identified above a number of additional benefits were derived. These largely fall in line with the general benefits identified from the successful application of e-learning methodologies as well as the development of LCA related skills.

1. Students can learn effectively from recorded lectures, by enriching traditional lecture/tutorial format. This can be due to the flexibility to view the recordings or videos at the times of their choosing and the ability to repeat sections of the recordings as needed;
2. Class time is efficiently used for students to work in small groups to answer content questions and discuss LCA case studies. This facilitates group dynamics and interaction during class time;
3. Teaching supported with recorded lectures, YouTube videos, online readings, embedded quizzes, and follow-up discussions can improve students' understanding and retention of LCA concepts;
4. Students use Blackboard to share messages and comments with the instructor and other students.

Developing an approach to address sustainability learning and advocacy

In developing an approach to the design of the courses we needed to fully appreciate both the theoretical concepts of e-learning and the objectives we are trying to achieve, especially with regard to supporting the next wave of professionals educated in the fundamental concepts of sustainable development. In order to do this, we firstly considered open learning environments and flipped learning methodologies as a base for the design of the model. The key themes for both concepts that were included were user centric learning, collaboration, problem-orientated learning and blended integration with physical lectures. A final aspect for consideration was a need to induce a sustainability focused mindset within learners. In taking these concepts into account we developed the e-learning advocacy model, the components of which are described in detail below.

Core Learning – The first consideration for design was to blend the traditional concepts of physical lectures and tutorials with online resources to teach students the fundamental concepts of what was being delivered for each weekly topic. The idea here is to provide one or two overview videos/resources that could then help leverage their initial understanding of sustainability concepts in class as well as allow student to effectively participate in discussion (Herreid and Schiller, 2013). In terms of theoretical concepts core learning is a fundamental component of self-directed-learning as students are able to acquire knowledge in their own time. Flipped learning and collaboration is also facilitated by the in-class discussions through both the lectures and tutorials.

Knowledge Development and Reinforcement – With an understanding of the core ideas, there was a need to include resources that could challenge their knowledge and help them adopt a deeper appreciation of the theoretical concepts. This component of the model encouraged supporting e-learning videos and resources that could go beyond the initial understanding and allow students to think more critically of the concepts involved. Once again this component focused on self-directed learning as students were able to personally guide what knowledge they looked to acquire.

Assessment and Engagement – Once a deeper understanding of the concepts were achieved, there is a need to utilise e-learning tools to help students test their knowledge and participate in online dialogue with both their peers and the instructor. This component was designed to allow students to apply what they have learnt through the use of assessment and interactive tools. The design also encouraged the use of online forums to go beyond the initial in-class discussion and participate in ongoing debate and critical analysis. An additional component related to the objectives would be the development of a dedicated LCA resource that could potentially include LCA Applications, datasets and guides. This component largely focuses around e-learning tools for both collaborative learning and problem-orientated learning. The intention here is to challenge the learner to reinforce a critical understanding of the concepts through collaborative engagement and support the acquisition of action competencies.

Advocacy and Real-World Application – A final component was the need to reinforce their support for environmental thinking and give students the tools and resources to help them apply concepts in the real world. This could be in the form of online resources for living sustainably, details on career pathways, additional study and learning options, social media tools as well links to additional environmental knowledge resources and initiatives. This component does not lend itself solely to one principle of open learning environments or flipped learning but is a consolidation of the ideas to help reinforce the overall context of the learning setting. The idea is to enhance their understanding of sustainable development beyond the realms of the course and give learners real life and alternative opportunities to apply their critical analysis skills and competencies.

Figure 1: e-learning Advocacy Model

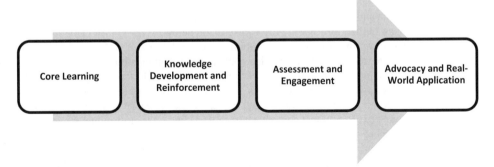

As an additional comment, in developing this model we must appreciate the fact that it had been initially tailored to satisfy the requirements of the project. As the model has adopted broader e-learning pedagogical concepts, it can be argued that the model may be relevant to other subject areas where an e-learning approach is required. Much of the model requires use of targeted thematic content applied in a logical and organised manner consistent with flipped learning and open learning environments, which in theory could be applied for any subject delivery. A discussion of the wider applications of this model is discussed towards the end of this chapter.

Applying the Model

The model and initial approaches were developed in the second half of 2013 for the delivery on ENVM2101 which commenced soon after. Content was designed so that the four components of the model were fully addressed and a comprehensive e-learning experience could be delivered. In terms of timescales, a budget of 60 hours for research and course development was allocated to both of the courses. The core course topics and learning objectives for the courses had previously been designed as part of the course profile, so this time budget was intended to apply flipped learning concepts and e-learning models to the Blackboard sites.

The first few hours of development focused on effective engagement between all relevant stakeholders involved in the project. The key individuals were; the course coordinator, who would design the course outline and deliver lecture material, course tutors who would support in teaching the theoretical concepts, and the e-learning teaching assistant who would focus on building the Blackboard content. Through this initial engagement the core components and high-level requirements were identified and developed. As the material on Blackboard was being built, a key effort was to maintain effective communication between stakeholders and ensure the course coordinator was happy in the content being able to reflect the sustainability learning he would like to achieve.

The Blackboard for the Sustainable Consumption and Production course was developed over a period of approximately three and a half months. Blackboard content was continuously being uploaded as the course was progressing. It was found that this approach facilitated the dynamic upload of content specifically related to each weekly topic. By also being regularly engaged with Blackboard it was also noted that academic stakeholders were able to maintain ongoing dialogue between themselves and students.

For Sustainable Consumption and Production, the Blackboard content was finalised a week or two in advance of the physical lecture. In additional to the weekly material for each topic, the Blackboard site was redesigned as a whole in order to support knowledge retention, student engagement and the adoption of a sustainability-focused mindset. A description of how the model was applied for ENVM2101 is discussion below.

Core Learning – As this focused on combining physical lectures with flipped learning strategies the main focus was to provide video content and reading materials that could leverage initial understanding for the lecture. For each weekly topic the lecture materials would be supported by one or two fundamental videos that students could watch to understand key sustainability concepts. Some of the videos that fell under this category included government insight videos, industry promotional content, short professional talks or introductory academic lectures. The idea here was to keep core learning videos to no more than 10-15 minutes in length and support this with fundamental readings. In addition, the e-learning materials were enhanced by the actual lecture recordings which were captured using Lectopia. This would help students to review the core concepts and recall discussion topics for later on in the semester. Figure 2 shows the structure of how the learning resources folder was presented. Our course design aimed to support easy navigation and clearly defined content areas.

Figure 2: Learning Resources Content Area

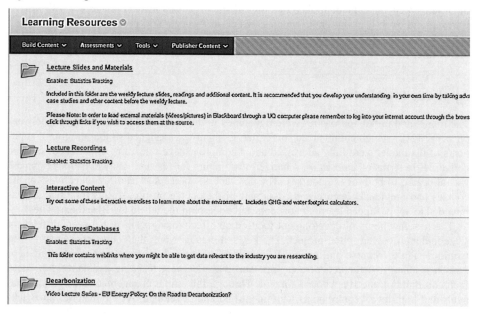

Knowledge Development and Reinforcement – The materials for this section had a focus on wider learning and developing critical thinking for relevant environmental concepts. A key video resource here was utilising TedX talks freely available online that offered new and interesting ways that could challenge the minds of the student audience. Videos of greater length were also included that would go into greater depth on environmental theory and included full academic lectures, documentaries, video series and case studies. This was in turn supported by specific environmental reports written by national governments and environmental bodies such as the United Nations Environment Programme (UNEP) and the Ellen MacArthur Foundation.

Also included were supplementary reading materials and academic journals focusing on related environmental ideas. Figure 3 shows how both core videos and supplementary videos were embedded directly into the Blackboard area. An additional recommendation is to add description to clearly identify the content and purpose of the video.

Figure 3: Embedded YouTube Videos

Assessment and Engagement – In order to put student's understanding into practice a weekly discussion question was posted on the Blackboard forum related to the specific sustainability/environmental concept. The aim here was not to be too specific but allow for open-ended answers with a particular focus on sustainability issues on a societal level. For the first few topics a weekly assessment quiz was also built that would allow students to test their knowledge. Due to the time in developing each of the quizzes it was decided to put on hold their inclusion and focus on testing their understanding in tutorials and marked assessment. The intention is to allow more time to develop multiple-choice questions with a focus on sustainability learning objectives for the IE course in early 2014. Figure 4 is a screenshot of the various discussion topics posted over time. These discussions helped facilitate peer engagement and collaboration.

Figure 4: Weekly Discussion Topic

▼ Date ▽	Thread
14/10/13 11:29	Promoting Pro-environmental behaviour on campus. What do you think are effective strategies to promote pro-environmental behaviour to students?
23/09/13 11:29	How can LCA help us better understand renewable energy? What do you think the general public feels about renewable energy? (e.g. can windfarms be an eyesore to some? do the public actually know enough about renewables?)
18/09/13 11:10	Do we consume too much food? What can be done to change our eating habits?
09/09/13 10:55	What is your understanding of LCA-Select an industry of your choice (e.g. food, tourism, mining etc) and quickly summarize how LCA can be applied. Your summary for the selected industry could potentially include the following: - Application - Internal benefits - External benefits - Limitations
27/08/13 13:25	What do you understand about product redesign? Discuss how these concepts have been applied to a particular product that you own.
19/08/13 10:16	Which clean energy technologies are you excited about? What do you think are the benefits and disadvantages of the differing technologies in Australia?

In addition, this component encourages interactive content to truly engage the student and set context. Examples of items for this section included carbon footprint calculators, water footprint calculators, interactive maps, smartphone applications and external links to self-assessment questionnaires and quizzes.

As part of the emphasis on Life Cycle Assessment this section also focused on giving hands on experience to specific LCA tools and a number of links to external applications was provided. To support a student's understanding of these tools, introductory videos highlighting functionality were also provided along with basic tutorial videos. In support of the LCA tools and applications were a number of real-life data sets that students could perform analysis on.

Advocacy and Real-World Application – Content for this section did not focus solely on delivering theory but on the how people can apply sustainable living. Specific sections within Blackboard were created that focused on sustainability initiatives, organisations and guides. Videos and articles focusing on items such as reducing your carbon footprint, greening the environment and recycling aimed to encourage students to put their knowledge into action. Encouraging students in their careers was seen as a way to reinforce a sustainable ideology and a dedicated section on Blackboard was built to aid this. The section included videos on careers in sustainability as well as links to job boards and careers resources.

A summary of the materials included for the Blackboard is identified in table 1.

Table 1:   Materials included for the Blackboard.

Core Learning	Knowledge Development and Reinforcement	Assessment and Engagement	Advocacy and Real World Application
Lecture materials and slides (1 PDF) Overview videos (1-2 YouTube videos) Core reading materials (2-3 papers or reports). Lectopia Lecture Recordings	Industry Videos (1-2) TEDx videos (1-2) Government and industry environmental reports/ factsheets Supplementary reading materials Case studies	Weekly discussion question utilising online forum Development of multiple choice quizzes LCA software packages Interactive Content	Sustainable living resources and guides Links to environmental resources, news and initiatives Career pathway videos Options for further study and environmental research Links to environmental careers websites

Outcomes

The overall results from the development of the Sustainable Consumption and Production Blackboard site as well as the current development for Industrial Ecology and Life Cycle Thinking were relatively positive. In broad terms there is a diverse range of materials available to support flipped learning strategies in teaching sustainability concepts. Online resources such as YouTube offer thousands of videos from a broad range of sources that can help support concept delivery. Blackboard also offers a number of tools to help facilitate flipped learning implementation and students were able to access content from both on and off campus. In terms of being able to use sustainability teaching to help support an environmental-mindset, there were plenty organisations, initiatives, career opportunities and guides to aid students long after they have finished the course. The following is a more detailed description of specific outcomes as well as how outcomes related to particular objectives.

One Stop Blackboard – In terms of matching the objectives with the outcomes it can be argued that the methodology utilised was relatively successful. The aim for a fully comprehensive Blackboard site was met due to the diverse variety of resources available and effective application of the e-learning model. Students were able to access case studies, videos, reading materials and public available sustainability resources that were all consolidated into one convenient location. Items such as YouTube videos were embedded directly into the site so that students could view sustainability content without the need to leave Blackboard.

Diverse video materials to support flipped learning/self-directed learning – The Internet is a wealth of knowledge and there are thousands of videos and resources dedicated to sustainability and LCA related concepts that can be uploaded onto Blackboard. Organisations such as UNEP, government bodies, corporations and individuals generate the content that can teach both the broad theory and specific ideas that a sustainability lecturer may want to deliver. However, it must be noted that some searches can become overwhelming with regard to what is relevant and concise enough for the students to learn from. Another challenge is that there may be difficult to obtain the exact video to help support the topic of the lecture. Overall video content did help support the course objectives and videos were viewed both online and even in class where appropriate.

Self-assessment – A challenge for self-assessment using multiple choice quizzes is being able to write numerous questions for each topic and keep them relevant, engaging and support understanding of sustainability theory. Another concern is the need to programme and test the questions on a weekly basis. There is capacity to develop and implement multiple-choice questions to help support those wishing to become certified practitioners in LCA. This is something that will be looked at in more depth for the specialised IE and Life Cycle Thinking course; however there is a positive outlook with regard to the broad spectrum of LCA related material online and the capacity for Blackboard to design multiple types of self-assessment. Another positive is the diverse range of related materials that can support self-assessment such as online quizzes, LCA case studies and model answers from past questions.

Student Engagement – Overall online student engagement was positive and a number of students did participate in discussions for topics in the early stages of the course. For the first discussion topic posted there were over 20 responses to the question posed however towards the end of the course participation dropped dramatically to only one or two replies. As a key factor in the underlying success of the model, maintaining effective student engagement over the duration of the course and understanding the drop in participation is something that warrants further investigation.

Diverse LCA Resources – Another positive were the resources available and tools developed to support and learn about Life Cycle Assessment. For those students wishing to become certified practitioners of LCA online tools such as OpenLCA and CMLCA were available to access and download. This enabled students to work with LCA tools on their relevant projects and potentially apply their understanding to other real-life scenarios. Within Blackboard a dedicated section was implemented which had over 30 links to tools and databases that students could utilise.

Promoting Sustainability – The course Blackboard provided a number of opportunities for student to adopt and engage sustainability as part of their everyday activities. Although it is difficult to gauge at this stage how effective they were at promoting sustainability it is more than likely that students are now more aware of what opportunities there are available and the specific approaches that can enable them to live sustainably. The dedicated Blackboard sections had links to 5 different careers resources as well as over 10 links to sustainability initiatives and resources.

Discussion and Conclusions

In this chapter we aimed to show how e-learning methodologies can be utilised to support course design and the delivery of theory related to sustainable industrial development. An overlying objective was to enhance the ability for students to apply critical analysis to related concepts and to support the development of competencies that they could utilise in their future careers. Of key interest were e-learning methodologies related to open learning environments and flipped learning. In particular, specific principles of self-directed learning, collaborative learning and problem orientated learning offered much potential in creating an effective learning environment and were utilised in the design of both courses.

In the application of the model and theoretical concepts the outcomes were overall fairly positive however the implementation was not without its challenges. What we found was that there was a significant amount of video content, reading materials and activities to support both self-directed learning and problem orientated learning. A potential challenge exists in filtering out the vast number of resources related to sustainable development so that material is both concise and informative. In contrast, despite these numerous resources it may be difficult to obtain the perfect video or material to support the learning of specific concepts. This may be where user-recorded videos may be most relevant. From a student perspective, the vast repository of information available does facilitate self-directed learning as they are free to explore and develop a critical understanding of topics that are of most interest to them. By utilising embedded content and targeted links in Blackboard a lecturer can provide guidance for topics, such as climate change, for a student who may be initially overwhelmed by what is available to them.

Collaborative learning outcomes and overall student engagement were both positive and negative. It seemed difficult to maintain student's interest in discussing topics over time. New and innovative ways may be required to encourage user engagement and shift students from the mentality of traditional passive recipients to active creator of knowledge insights. It must be noted that this was also a challenge identified by Barth and Burandt (2013) who suggested as systematic way to encourage learner confidence and enhance skills may be needed. Whilst it is difficult to pinpoint the exact reasons for the reduced level of student contribution over time in this particular case, broader research by Hew and Cheung (2012) offer a multitude of different reasons as to why online participation is often relatively low amongst learners. Potential reasons that may be relevant to the course may include; not seeing the need for online discussion, not knowing what to contribute/lack of worthwhile comments to contribute, student personality traits, displaying low-level knowledge construction and lack of time (Hew and Cheung, 2012). Being an introductory course could mean that students have not developed the theoretical understanding (in their minds) to form valuable opinions that may be relevant to others. Another potential reason is that over time some of the discussion questions moved from generic open-ended sustainability topics to those with a focus on more technical LCA concepts. This may initially be a challenge for those who are still acquiring the theoretical knowledge to put into practice.

Finally another explanation could be that as student's reading and coursework pressures build up the duration of the semester they simply lack the time to engage in online discussion. Hew and Cheung (2012) acknowledge the fact that student participation rates is a broad topic for research in itself, however what can be taken from this study is that formal methodologies to gauge student opinions and strategies to address student/teacher engagement may be required. This paper looks to initially explore these options in the discussion section below.

In terms of measuring direct impacts on student's learning, which may be hard at this time, but what can be said is that the objective to create a one-stop Blackboard was, by and large, mostly achieved. The capacity for e-learning tools to create a consolidated and accessible resource for sustainable development learning is great and this insight can be used in future course applications. It must be also noted that e-learning does facilitate new and innovative ways to support course design and can potentially enhance the delivery of education for sustainability. This potential can only be realised with ongoing applications of e-learning theory to support course design, along with an underlying objective of educating individuals who truly appreciate and are able to apply sustainable development concepts.

Further Discussion: Refining the e-learning Advocacy Model

With the model having being developed and initially tested with the Sustainable Consumption and Production course we can use our findings to refine the model and improve outcomes for future applications. This is particularly relevant to the upcoming Industrial Ecology and Life Cycle Thinking course. Through using the modelling approach to course design two development areas emerged that have potential to add value to the model and increase learning success. These are (1) creating preliminary content to enhance initial stakeholder buy-in whilst incorporating strategies to maintain ongoing student participation; and (2) the development of further methodologies to systematically analyse student's learning experiences and impacts on the overall acquisition of knowledge.

As a pilot implementation to test the model, the project did not overtly express to learners at the start of the course that a new approach to learning was being trailed. The project looked to adopt staged e-learning whilst incorporating learning design which had been successful for the previous iterations of the course. For future applications of the model it may be beneficial to inform students at the start of the course of what is expected and what can be achieved from online participation. Hew and Cheung (2012) noted students not seeing the need for online discussion as a factor in lower levels of student interaction. Recommendations gathered by Hew and Cheung (2012) from other empirical based studies include providing a clear explanation to students of instructor expectations and the principles of online discussion (Jung et al. 2002; Yeh and Buskirk, 2005). With this in mind an additional component can potentially be added to the model that involves using short videos or instructor notes to explain e-learning objectives and the benefits that can be achieved through collaborative learning amongst participants. One of the aims that can also be identified early on is to inform students of the long term course goals of creating a sustainability focused mindset and the diverse resources available online that can help facilitate this. Delivering expectations for the course is not only a one way exercise, it may also be advantageous to understand the initial e-learning expectations of students who are about to start. Paechter, Maier and Macher (2009) identified the fact that discovering the achievement goals of students' can contribute to the overall success of e-learning outcomes. A recommendation here would be to develop strategies, such as online forms and questionnaires, to obtain insight from those who are about to undertake the course and tailor content to address this. For example greater focus on renewable energy sources or product redesign can be implemented for students who show strong interest in developing expertise in these areas.

Whilst it is important to create an initial base for e-learning support, it is also a key to build on this throughout the course and use the multitude of different strategies that can facilitate ongoing online engagement (Hew and Cheung, 2012). Tutor involvement in the discussion is one method to improve weak participation amongst student learners (Hew and Cheung, 2012; Tagg and Dickinson, 1995) and the course looked to address this through educators setting discussion questions, voluntarily providing feedback through online communications and tutor engagement in physical tutorials. However a potential recommendation may be to hire additional teaching assistants whose purpose is to not only take part in physical tutorials but who are required as part of their contract to engage in regular discussion (e.g. at least 1 hour over the course of the week). Rovai, Ponting and Wighting (2007) noted that graduate students exhibited higher intrinsic motivation for e-learning and traditional classroom based approaches than undergraduates. In the context of our course, although it is designed for both undergraduates and graduates, a suggestion may be to hire volunteer graduate students who have already passed the course or who have advanced expertise in related sustainability subjects to act as e-learning engagement ambassadors to participate in discussion.

The recommendations above are ones that allow for easy implementation and fluid transition to future applications of the model, however there are still a diverse range of additional strategies that have scope to be explored. One that can be potentially facilitated by Blackboard and allows for learner driven content development is the use of wiki technologies. Giannoukos et al. (2008) found that using such technologies helped support teamwork amongst students, enhanced learner focus and growth and led to the creation of a satisfactory level of educational content. In the context of sustainability education, Barth and Burandt (2013) found that collaborative wikis acted as effective knowledge management systems in their "*Interdisciplinary Study Program on Sustainability*" case study. For future applications of UQ's sustainability education courses, a student developed wiki platform can be created within Blackboard to store materials based on sustainability discussions assigned by tutors. Other potential developments that are worth investigating may include the use of mandatory grading of discussions to encourage student input and peer-facilitation whereby the learners themselves are encouraged to instigate discussion (Hew and Cheung, 2012). The case for using marking and grading of collaborative inputs is an interesting one as it offers additional dimensions for student assessment which have not yet been applied in previous course iterations. It must be said however that with the multitude of different e-learning approaches to enhancing student engagement, it may be difficult to fully explore all options at this time and investigating these ideas is both an interesting and challenging project for future research in online sustainability education.

Whilst improving student engagement is a key aspect of the discussion, another consideration that may leverage learning outcomes is creation of insight through evaluation of student perceptions and study experiences. In this application of the model we did not explicitly set out to obtain empirical evidence from students. However, informal measures of student perceptions were analysed through monitoring online Blackboard discussions and activity. Now that we know that about the availability of content and Blackboard tools to support sustainability e-learning, a holistic methodology to gauge overall effectiveness from a learner perspective can now be designed. Applying traditional research methodologies to e-learning courses is something that offers an excellent starting point and there are a number of different empirical approaches available focusing on student satisfaction and outcomes in online environments (Eom and Arbaugh, 2011). In their recent study on how university students are impacted by e-learning systems, Alkhalaf et al. (2012) utilised the IS Success/Impact Measurement framework (DeLone and McLean, 1992; Gable, Sedera, and Chan, 2008) to discover positive impacts in a number of learning areas including student productivity, student satisfaction and depth of learning. Whilst statistical methodologies and highly developed IS learning frameworks may be very useful for a future research project based solely on measuring students' outcomes, in the context of this project, an excellent future application would be to utilise tools available in Blackboard to gauge e-learning perceptions and results. Blackboard offers the potential for questionnaires, forums and journals to be tailored in order to uncover how students are developing their understanding, how they perceive the effectiveness of the course and how their attitudes towards sustainability are changing. In order to do this a structured approach is required with measurements being gathered at different stages of course delivery. The potential approaches to acquiring insight at the commencement stage, the course delivery stage and the post study stage are analysed below.

As mentioned previously, gathering student's expectations and learning objectives prior to commencing the course may offer opportunities to discover where their interest lies and what areas of sustainability to focus discussion topics on. A simple questionnaire may be the most feasible option as it offers an easy way for students with limited understanding on sustainability concepts to provide feedback without having to explain their beliefs in too much detail. Another option may be to create polls that students can vote on, which again allows them to easily express their beliefs on sustainability issues without them being challenged to defend them. As the course progresses a tool that has potential to obtain in-depth qualitative feedback is the use of journals that students can fill in to explain how well their learning is progressing and how effective e-learning approaches are in their acquisition of knowledge. It may also be useful to have an ongoing discussion topic that focuses on student's e-learning opinions and allows them to discuss their ideas amongst peers.

Encouraging students to adopt a sustainability mindset is one of the pillars that support the e-learning Advocacy Model and understanding changes in their belief system on course completion is something that will determine its overall success. Qualitative results in the form of short answer questionnaires developed in Blackboard as well as opinion polls offer great potential to gauge opinions. An interesting option that focuses on student generated content is the potential to create videos that can be posted to the Blackboard site. Videos offer a more personal touch and can allow students to personally explore how their opinions and beliefs around sustainability concepts have developed during the course of their study. In the long term a recommendation may be to conduct short face to face interviews with students who are about to graduate from their programmes of study. Whilst this moves away from the technological capabilities of Blackboard, it does allow for in depth study to really delve into how a student's attitudes to sustainability were influenced by the courses and how their perceived level of competency in areas such as LCA have developed.

What can be shown by effective investigation of experiences, behaviours and learning impacts is useful insight that can affect decisions on course design and e-learning approaches (Alkhalaf et al. 2012; Hew and Cheung 2012). Insights gained from Barth and Burandt's (2013) study on sustainable education course deliveries include the fact that e-learning tools offer further opportunities for reflective actions, effective platforms for knowledge management and facilitation of discussion across intercultural dimensions. What was also noted, however, was that in some cases asynchronous discussion can be time consuming and the value of face to face contact through physical engagement and nonverbal communication cannot be underestimated (Barth and Burandt, 2013). It is hoped that by applying a staged approach to gaining feedback that we can acquire similar insights which can be applied over the duration of the course. A blended learning approach still appears to be the best solution to enhanced learner competency in sustainability concepts and fully understanding what e-learning strategies complement traditional pedagogies will significantly help leverage future course deliveries.

One final comment regarding the model is its suitability for application in other subject areas. In developing the model an analysis of current literature on technologies to deliver sustainability education was undertaken. It was soon discovered, however, that insight in this specific area was in its relative infancy and so e-learning theory from other subject areas had to be incorporated in order to justify the new approaches. Concepts that underpin the e-learning advocacy model such as flipped learning are not unique to sustainability education. We do not attempt to redefine e-learning theory, rather our findings show that the model offers some level of guidance with organising relevant materials and exhibiting how learner centric, collaborative approaches can be used in the context of e-learning for sustainability and LCA education. It can therefore be argued that some generic e-learning concepts in the model could potentially be applied to teaching other subject areas if the material is tailored to the needs of the specific course. An example of the multidisciplinary nature of the model, and e-learning theories in general, are the case studies presented by Barth and Burandt (2013) where courses use collaboration tools such as file sharing and forums to help support sustainability education across higher level institutions. These concepts are identified by Hew and Cheung (2012) as some of the tools that can support participation amongst learners across a broad spectrum of subject areas.

Following on from e-learning concepts and tools, we must acknowledge the fact that Blackboard in itself is not specifically designed for sustainability education or any other particular subject area for that matter. It is up to educators to design their courses and make use of the tools in Blackboard to help leverage their academic objectives in any way they see appropriate. What we have also done is highlight the possibilities that Blackboard offers in facilitating sustainability related collaboration and discussion, self-directed learning through structured materials and access to wider LCA/sustainability resources available online.

A final consideration is that if we view the model from outside the technical realm of e-learning and Blackboard, we can observe that it also incorporates wider concepts in sustainability education that aim to leverage behaviours that promote sustainable development in the long term. Approaches such experiential-learning (Brundiers et al., 2010) and problem-based learning (Brundiers and Wiek, 2013; Thomas, 2009) helped to influence the model and are both present in current pedagogical approaches to education for sustainability (Barth and Burandt, 2013; Barth and Michelsen, 2013). Once again it can be noted that these concepts are not entirely unique to sustainable development education and approaches such as problem based learning have been successfully applied in other academic fields (Schwartz, 2013).

In summary of the above points a theme that emerges is that the model builds on conceptual foundations that have been applied both to sustainable development education and other fields of study. Through the use of targeted sustainability content and approaches to induce pro-environmental behaviours, the model has been tailored to meet the specific needs of our courses. Although the current modelling approach may require further refinements, it can be argued that if one were to adapt the thematic content there is no reason why the model could not be utilised as a starting point for different subjects with an aim adopt e-learning concepts. This may be particularly useful in related environmental fields or social courses where the long term objectives may be to facilitate ongoing collaboration and discussion or induce particular behaviours or ethical approaches.

Acknowledgments An acknowledgments must be made to the authors Matthias Barth and Simon Burandt whose 2013 paper on "*Adding the 'e-' to Learning for Sustainable Development: Challenges and Innovation*" helped guide the development of this case study.

References

Alkhalaf, S., Drew, S., & Alhussain, T. (2012). Assessing the impact of e-learning systems on learners: a survey study in the KSA. *Procedia-Social and Behavioural Sciences* 47, 98-104.

Barth M, and Burandt S. (2013) Adding the "e-" to Learning for Sustainable Development: Challenges and Innovation. *Sustainability* 5(6), 2609-2622.

Barth, M., & Michelsen, G. (2013). Learning for change: an educational contribution to sustainability science. *Sustainability Science* 8(1), 103-119.

Bielefeldt, Angela R. (2013) "Pedagogies to Achieve Sustainability Learning Outcomes in Civil and Environmental Engineering Students." *Sustainability* 5(10), 4479-4501.

Brundiers, K.; Wiek, A.; Redman, C.L. (2010) Real-world learning opportunities in sustainability: from classroom into the real world. *International Journal of Sustainability in Higher Education* 11, 308-324.

Castells, M. (2011). The rise of the network society: The information age: Economy, society, and culture (Vol. 1). Wiley.com.

Daniel, J. (2012). Making sense of MOOCs: Musings in a maze of myth, paradox and possibility. *Journal of Interactive Media in Education, 3.*

DeLone, W., & McLean, E. (1992). Information systems success: The quest for the dependent variable. *Information systems research* 3(1), 60, 95.

Eom, S. B., & Arbaugh, J. B. (Eds.). (2011). Student satisfaction and learning outcomes in e-learning: an introduction to empirical research. Information Science Reference.

Gable, G., Sedera, D., & Chan, T. (2008). Re-conceptualizing information system success : the IS-Impact Measurement Model. *Journal of the Association for Information Systems,* 9(7), 377-408.

Geography, Planning and Environmental Management, School of, (GPEM) (2013) Available online: *http://www.gpem.uq.edu.au/school-profile* (accessed on 5 November 2013).

Giannoukos, I., Lykourentzou, I., Mpardis, G., Nikolopoulos, V., Loumos, V., & Kayafas, E. (2008, July). Collaborative e-learning environments enhanced by wiki technologies. In Proceedings of the 1st international conference on Pervasive Technologies Related to Assistive Environments (p. 59). ACM.

Herreid, C. F., & Schiller, N. A. (2013). Case studies and the flipped classroom. *Journal of College Science Teaching* 42(5), 62-66.

Hew, K. F., & Cheung, W. S. (2012). Student Participation in Online Discussions (pp. 16-25). Springer, Berlin.

Janz, K., Graetz, K., & Kjorlien, C. (2012, October). Building collaborative technology learning environments. In *Proceedings of the ACM SIGUCCS 40th annual conference on Special interest group on university and college computing services* (pp. 121-126). ACM.

Jung, I., Choi, S., Lim, C., & Leem, J. (2002). Effects of different types of interaction on learning achievement, satisfaction and participation in web-based instruction. *Innovations in Education and Teaching International* 39(2), 153-162.

Katz, J.E.; Rice, R.E. (2002) Social Consequences of Internet Use: Access, Involvement, and Interaction; MIT Press: Cambridge, MA, USA.

Knewton. (2012) The Flipped Classroom Infographic. *http://www.knewton.com/flipped-classroom/*

Mossberger, K.; Tolbert, C.J.; McNeal, R.S. (2008) Digital Citizenship: The Internet, Society, and Participation; MIT Press: Cambridge, MA, USA.

McNamara, K. H. (2010). Fostering sustainability in higher education: a mixed-methods study of transformative leadership and change strategies. *Environmental Practice*, 12(01), 48-58.

National Council for Science and the Environment NCSE (2003).Recommendation for Education for a Sustainable and Secure Future: A Report of the Third National Conference on Science, Policy and the Environment, NCSE, Washington, DC, 87pp. Available at *http://www.cnie.org/NCSEconference/2003conference/2003report.pdf.*

Paechter, M., Maier, B., & Macher, D. (2010). Students' expectations of, and experiences in e-learning: Their relation to learning achievements and course satisfaction. *Computers & Education* 54(1), 222-229.

Rovai, A., Ponton, M., Wighting, M., & Baker, J. (2007). A comparative analysis of student motivation in traditional classroom and e-learning courses. *International Journal on E-learning* 6(3), 413-432.

Schwartz, P. (2013). Problem-based learning. Routledge, London.

Sterling, S. (2001) Sustainable Education: Re-visioning Learning and Change; Green Books Ltd.: Foxhole, UK.

Stewart, M. (2010). Transforming higher education: a practical plan for integrating sustainability education into the student experience. *Journal of Sustainability Education* 1, 195-203.

Strayer, J. F. (2007). The effects of the classroom flip on the learning environment: a comparison of learning activity in a traditional classroom and a flip classroom that used an intelligent tutoring system (Doctoral dissertation, The Ohio State University).

Szafir, D., & Mutlu, B. (2013, April). ARTFul: adaptive review technology for flipped learning. In *Proceedings of the 2013 ACM annual conference on Human factors in computing systems* (pp. 1001-1010). ACM

Tagg, A. C., & Dickinson, J. A. (1995). Tutor messaging and its effectiveness in encouraging student participation on computer conferences. Journal of Distance Education, 10(2). Accessed 24th February 2014 from *http://www.jofde.ca/index.php/jde/article/view/238/599*.

Thomas, I. (2009) Critical Thinking, transformative learning, sustainable education, and problem-based learning in universities. *J. Transform. Educ.* 2009, 7, 245-264.

Tucker, B. (2012). The flipped classroom. *Education Next, 12*(1), 82-83.

United Nations Department of Economic and Social Affairs (UN DESA) (2008) 'Addressing climate change in national sustainable development strategies – common practices', Commission on Sustainable Development, Sixteenth Session, 5-16 May 2008, New York. Available Online: *http://www.un.org/esa/sustdev/csd/csd16/documents/bp12_2008.pdf* (accessed on 5 November 2013).

United Nations Educational, Scientific and Cultural Organization (UNESCO). (2009) Bonn Declaration, UNESCO World Conference on Education for Sustainable Development, Bonn, Germany, 31 March to 2 April 2009. Available online: *http://www.esd-world-conference-2009.org/fileadmin/download/ESD2009_BonnDeclaration080409.pdf* (accessed on 5 November 2013).

Vare, P.; Scott, W. (2007) Learning for a change: Exploring the relationship between education and sustainable development. *J. Educ. Sustain. Dev.* 1, 191-198.

Walton, A., Weller, M., & Conole, G. (2008). SocialLearn–Widening Participation and Sustainability of Higher Education. *Distance and E-Learning in Transition* 691-700.

Yeh, H. T., & Buskirk, E. V. (2005). An instructor's methods of facilitating students' participation in asynchronous online discussion. In C. Crawford, D. A. Willis, R. Carlsen, I. Gibson, K. McFerrin, J. Price, and R. Weber (Eds.), *Proceedings of Society for Information Technology and Teacher Education International Conference 2005* (pp. 682-688). Chesapeake, VA: AACE.

Greening Higher Education qualification programmes with online learning

Sally Caird, Andy Lane and Ed Swithenby[1]

Abstract

Digital infrastructure and devices and computer technology-based pedagogical applications are transforming the way higher education (HE) teaching, learning and assessment is delivered, and are having varied environmental, pedagogical and economic impacts. This chapter introduces the SusTEACH Modelling Tool, developed following an investigation of the impact of computing technologies or Information and Communication Technologies (ICTs) on HE teaching models together with a carbon-based environmental assessment of 30 courses, offered by 15 UK-based HE institutions. This offers lecturers a tool for modelling UK-based course and qualification programmes, and estimating their energy consumption and CO_2 emissions, developed following an analysis of the relationship between various online and ICT-enhanced, face-to-face and print-based distance teaching models and the main sources of energy consumption in HE. Applied to the Open University's BSc Environmental Management and Technology programme, which uses various Online and ICT-enhanced Distance teaching models, the Tool shows that compared with a wholly face-to-face taught programme, carbon reductions of 84% are achieved. Discussion includes the role of online learning designs and pedagogical use of computer technologies for achieving carbon reduction in HE.

Introduction

This chapter is concerned with environmental sustainability and the question of whether online learning designs and the pedagogical use of computer technologies deliver better or worse environmental impacts than other teaching delivery designs. It describes the Sus-TEACH Modelling Tool, which was developed at the Open University (OU) following research investigations and analysis of findings from the SusTEACH project, which is an acronym that stands for **Sus**tainability **T**ools for **E**nvironmental **A**ppraisal of **C**arbon impacts of **H**igher education teaching models using ICTs (see *http://www.open.ac.uk/blogs/susteach/ ?page_id=2*) . It describes how this tool was applied to estimate the energy consumption and carbon impacts in a new case study of the OU's BSc (Honours) Environmental Management and Technology qualification programme.

In recent years, UK higher education (HE) has been transformed by the use of computer technologies or Information and Communication Technologies (ICTs) which refer to digital resources and technologies utilised for the preparation, administration, and provision of teaching, learning and assessment, and the infrastructure, such as Virtual Learning Environments (VLEs), networks and cloud computing services that supports this provision. The Open University (OU), as a leader in the provision of distance education and supported open learning,

---

1   The Open University, Milton Keynes, MK7 6AA, UK.

has been at the forefront of establishing ICT-based infrastructure, including the equipment and networks that support platforms housing educational content, tools and applications within learning systems. Developing this infrastructure has enabled a transition from print-based teaching delivery systems to new online or e-learning designs.

In the OU, as well as more widely in HE, ongoing experimentation seeks to address the challenges and opportunities offered by computer technologies and online multi-media to deliver high quality, large-scale, accessible, cost-effective and sustainable teaching and learning. Computer technologies allow institutions to offer online or e-learning experiences that address requirements for temporal and spatial flexibility in terms of when and where students learn, as well as individualised learning designs in terms of how they learn. Computer technologies also support online interactive learning experiences that involve collaborative working and the creation of 'collective intelligence' that is valued by employers, and is consequently also driving pedagogical innovation (Johnson et al., 2012, p8).

Experimental use of hardware and software within the HE infrastructure has supported technology-enhanced teaching and learning provision (see *www.jisc.ac.uk*). It has also favoured pedagogical innovation, such as the development of digital education resources. This includes both 'closed', fully copyrighted resources, and 'open' openly-licensed online open educational resources (OER) (such as found on OpenLearn *www.open.edu/openlearn/* from the OU), and wholly online elearning courses and qualification programmes. This has led to radical new online learning designs, including the massification of education as a result of collaborative university partnerships to provide Massive Open Online Courses (MOOCs). These have widened access to global online learning communities. Examples include the Futurelearn platform led by the OU – *www.futurelearn.com/*, Coursera *www.coursera.org/* and Udacity *www.udacity.com/* set up by Stanford University, and Edx set up by Harvard University and Massachusetts Institute of Technology (MIT), *www.edx.org/about*). Some of these initiatives have a role in encouraging students to become both producers and users of educational resources (Lane, 2010).

Pedagogical applications emerging from computer technologies are regularly reviewed in reports, such as the annual New Medium Consortium Horizon reports (Johnson et al., 2012), and the Open University (OU) reports on Innovating Pedagogy (Sharples et al., 2012, 2013). These reports highlight the challenge for HE institutions to radically rethink teaching and learning designs, and the way that these are delivered, supported and assessed (see Conole, 2013). They also raise the question of what is a sustainable HE teaching model, in terms of pedagogical, economic, and environmental criteria.

Few studies have examined the environmental impacts of HE systems of delivering teaching and learning. One exception, the Factor 10 Visions study 'Towards Sustainable Higher Education' examined the key sources of energy consumption and carbon emissions of campus-based versus distance learning HE delivery systems (Roy *et al.*, 2005). Building on this earlier work, the SusTEACH research and development project conducted a carbon-based environmental assessment of 30 UK HE courses in campus-based (19) and distance-based education systems (11) from 15 HE institutions. The institutions employed different pedagogical designs for delivering teaching, learning and assessment, including online, ICT-enhanced, and traditional face-to-face and print-based teaching methods for course provision (Caird et al., 2013). For clarity, 'course' or 'module' are terms used in HE to refer to the set of modular, standardised, independent, or interrelated teaching units that comprise an undergraduate or post-graduate qualification programme. Degree programmes may also be called 'courses' although to avoid confusion, the term course is used in the first sense to include component courses and modules, rather than qualifications.

The SusTEACH research findings have been reported elsewhere (Caird et al., 2013) and are summarised here to provide essential background information on the steps taken to develop the SusTEACH Toolkit, which followed the analysis of the relationship between teaching models used in different HE institutions and the energy consumption and CO_2 emissions of courses. The toolkit includes the online SusTEACH Planning Tool (*http://www9.open.ac.uk/susteach*) which is a quick tool for lecturers and academic designers to use to consider their proposed design for a new or existing course and to obtain feedback based on the likely carbon impacts associated with this plan. Whilst the Planning Tool is helpful to begin with, the SusTEACH Modelling Tool *http://www9.open.ac.uk/susteach/background.htm#downloads* offers lecturers and academic designers an Excel-based tool to allow detailed modelling of HE courses and qualification programmes, as discussed in this chapter.

Background to developing the SusTEACH Modelling Tool

To examine the energy consumption and carbon impacts of courses, an attempt was made to identify, classify and compare courses using different teaching models for the planned teaching, learning and assessment provision, and the associated involvement of computer technologies in pedagogical design and delivery. This led to the development of a Teaching Models Rating Tool which enabled lecturers to rate how they planned to use different teaching delivery methods for course provision. The methods included face-to-face teaching, print-based educational materials, and computer technologies and rich media to supplement or replace traditional methods (Caird and Lane, 2013). This tool was further developed online (for example *http://www9.open.ac.uk/susteach*).

Based on lecturers' ratings, this approach permitted the classification of each course based on its primary HE teaching model, using the following classification of HE Teaching Models:

- The Face-to-face Teaching Model uses mainly face-to-face teaching methods with no enhancement using computing technologies;
- The ICT–enhanced Face-to-face Teaching Model uses face-to-face teaching methods enhanced by the use of computer technologies, e.g. to provide online links to downloadable resources;
- The Distance Teaching Model uses mainly classic distance teaching methods such as using printed educational materials with supported learning and which has little or no enhancement using computing technologies;
- The ICT–enhanced Distance Teaching Model uses classic distance teaching methods, enhanced by some use of computer technologies e.g. to provide online links to downloadable resources or audio-visual digital resources;
- The Online Teaching Model provides mainly online teaching, learning and assessment, available via the course/module Virtual Learning Environment. The model may include minimal printed materials, and small amounts of face-to-face teaching e.g. to attend day schools.

This classification was used to support the SusTEACH research analysis following an environmental assessment of the 30 UK HE courses, noted above (Caird et al., 2013). The SusTEACH Modelling Tool rationalises and synthesises data associated with the various teaching model designs noted above. This data includes the results of the analysis of the average energy consumption and CO_2 emissions associated with HE courses, which was based on data collected from students, lecturers and HE institutions, using online questionnaire surveys about course-related activities, and additional data gathered from databases, estates data and modelling software. These data sources included:

- Travel to and from places where teaching or learning takes place;
- Purchase and use of ICT devices;
- Purchase of books and publications, the provision of educational materials, and the use of paper for printing and photocopying;
- Residential energy use by students and lecturers;
- Campus site operations, including specific data collected for HE distance teaching systems on course production and presentation, and transportation of teaching materials.

The results of student and staff course-related activities were normalised using the standard UK Credit Accumulation and Transfer Scheme (CATS) system of HE institutional arrangements for measuring student progression towards defined learning outcomes and qualifications. This is a time-based measure for comparing the impacts of courses with defined study hours (*www.qaa.ac.uk*). This partly matches the European Credit Transfer Scheme within the European HE Area (ECTS, 2009). The CATS system identifies 1 CATS credit as equivalent to 10 hours total study. Study includes writing assignments, field work, etc. In summary, 120 CAT credits is equivalent to one student's full-time study per academic year; 360 CATS credits are required for an UK undergraduate Bachelor's degree; 180 credits for a post-graduate Master's degree.

The normalised course activity data was converted into energy consumption in megajoules (MJ) and CO_2 emissions in kilograms of carbon (Kg CO_2) using the latest carbon conversion factors available from the UK Departments for Environment, Food and Rural Affairs (Defra) and Energy and Climate Change (DECC). These provide conversion factors for all fuel sources, based on units of consumption and transportation, in a UK context (Defra/DECC, 2011). In addition, measures of embodied energy were gathered from environmental impact life-cycle studies to provide measures for paper, printed materials, and computer technology equipment (Caird et al. 2013). They include calculations of primary energy consumed over the life-cycle of a product or system associated with extraction, production, distribution, use and eventual disposal that gives rise to indirect emissions. These data were averaged per student, in order to derive the average energy consumption, and CO_2 emissions of a course, using a measure per student/ per 10 CATS credits (equivalent to 100 hours of study) (Caird et al., 2013).

In addition to modelling the impacts of different teaching model designs, the Modelling Tool (figure 1) includes an assessment of different energy impacts as a result of student travel behaviour and consumption of materials. For example, there may be considerable variation in transport impacts due to the impact of student air travel between the home and term-time residence of non-UK domiciled students. The tool accounts for these differences, and also allows the modelling of full qualification programmes that include a number of courses with similar or different delivery models.

Figure 1: The SusTEACH Modelling Tool

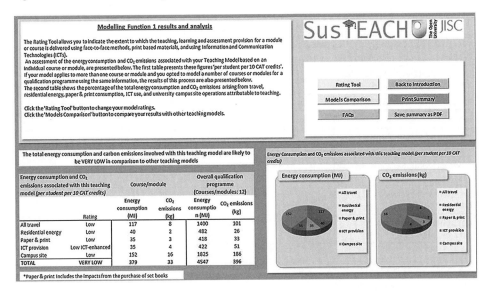

Application of the SusTEACH Modelling Tool to the BSc qualification programme

This SusTEACH Tool was employed to estimate the likely energy impacts of the BSc (Honours) Environmental Management and Technology qualification which was offered by the OU in 2013 *http://www3.open.ac.uk/study/undergraduate/qualification/q72.htm*.

The BSc programme has 3 stages, equivalent to 3 years in a full-time, campus-based system, covering 120 study credits per year/stage. The delivery is flexible in order to allow students to study at varying paces to attain their degree over several years. Academic lecturers who were teaching on the component courses of the BSc programme were asked to identify the teaching methods used in their course(s). The BSc programme provision adopts a mixture of online, ICT-enhanced, print-based distance and face-to-face teaching methods to provide 8 courses equivalent to 360 CATs credits. Such blended or hybrid models that combine online and traditional teaching and learning are predicted to become the dominant scenario in HE (see Johnson et al. 2012). Table 1 presents an overview of this qualification programme in 2013 with a summary of teaching methods.

Table 1:   Overview of OU's BSc (Honours) Environmental Management and Technology qualification programme.

Course (Code)	Study credits (CATS)	Face-to face teaching	Provision of printed educational materials	ICT-enhancement and online teaching
Stage 1				
Environment: journeys through a changing world (U116)	60	Low	High	High
Exploring science (S104)	60	Low	Medium-High	Medium
Stage 2				
Energy and sustainability (T213)	30	Low	Medium-Low	High
Environmental management 1 (T219)	30	Low	Low	Online
Environmental science (S216) OR	60	Low	High	Medium
Environment: sharing a dynamic planet (DST206)		Low	High	Medium
Stage 3				
Renewable energy (T313)	30	Low	Medium-Low	High
Environmental management 2 (T319)	30	Low	Low	Online
Innovation: designing for a sustainable future (T307)	60	Low	Medium-High	Medium

Modelling face-to-face teaching

All courses in this qualification programme had low levels of face-to-face teaching, although most included several contact classes (day schools) held in study centres (e.g. U116, S104, DST206 and T307). All the BSc courses require low amounts of travel during the course of study. Students typically live at home during the programme, and only travel to a study site (e.g. campus or study centre) a few times during the programme.

The SusTEACH Tool models the likely impacts of face-to-face teaching on energy consumption and carbon emissions by examining the impacts of course-related travel and accommodation. The impacts of term-time travel using guided rating options, and if applicable, travel between home and student accommodation and residential energy consumption are similarly assessed. Although not applicable in this case study, the SusTEACH Tool may also be used to model courses where students live in a university or temporary residence during their studies and need travel to and from home at the beginning and end of term or semester. Thus, the Tool captures this category of campus-based course which includes higher proportions of overseas students.

Modelling provision of printed educational materials

Although all the BSc courses provide printed educational materials, there were differences in the amount of printed materials used for the course provision. Some courses provide printed course books (e.g. U116, S216, S104, DST206), a set text (e.g. T213, T313), or pdf alternatives to print (e.g. T307), while others provide minimal print, with a course reference book, while being mainly online (e.g. T219, T319).

In this case study, the SusTEACH Tool identified two ratings for the use of printed educational materials. The HIGH RATING option applied when the majority of teaching was delivered using printed materials, in the form of paper, books, and other publications provided to students that were equivalent or greater than 250 sheets/pages per 10 CAT credits (e.g. U116, S104, S216, DST206, and T307). The LOW RATING option was found to apply when the main teaching method was not print-based, and any printed material provided to students was likely to be less than 250 sheets/pages per student per 10 CAT credits (e.g. T213, T219, T313, and T319). In addition to the materials provided, students may be required in some courses to purchase additional publications, which the SusTEACH Tool is capable of modelling, although this did not apply to any of the BSc courses under investigation.

Modelling ICT-enhanced and Online teaching, learning and assessment

All of the BSc courses were offered through the OU's VLE, although each course varied in the extent to which they were enhanced by computer technologies or online activities.

The SusTEACH Tool identified a HIGH RATING teaching option that applies when teaching, learning and assessment is provided mainly online, using computer technologies and digital resources available on the university websites and VLE. For example in this study, the BSc courses T219 and T319 are mainly online, although they have some minimal use of printed materials.

The ICT-enhanced teaching HIGH RATING option applies when the provision is strongly enhanced by computer technologies, for example via online links to downloadable resources, or offline using specially produced audio-visual digital resources (e.g. U116 offers DVDs and online activities. T213 and T313 have audio podcasts, online course materials, and some interactive computer-marked assessment (iCMA).

The ICT-enhanced teaching MEDIUM RATING option applies when there is some limited enhancement by computer technologies. (e.g. S216 offers virtual field trips, and three offline DVDs with interactive activities, although there are not many online links to resources and there is no need for students to use computer devices for parts of the course. S104 has some DVDs and some iCMA. T307 offers downloadable resources, online tutorials and forums. DST206 offers a course website, online forums and some iCMA.)

The ICT-enhanced teaching LOW RATING option applies when the provision is mainly provided using face-to-face or print-based teaching materials and has little or no enhancement by computer technologies. This was not applicable to any of the BSc courses under investigation.

In summary, the SusTEACH Tool was used to estimate the energy impacts for the BSc courses, which were grouped together according to their teaching model design.

1. Courses with the ratings: Low face-to-face teaching; High printed materials; and Medium ICT-enhanced teaching [including S104 (60 CATS), T307 (60 CATS), S216/DST206 (60 CATS)] provide 180 CATS study credits equivalent.
2. Courses with the ratings: Low face-to-face teaching; High printed materials; and High ICT-enhanced teaching [including U116 (60 CATS)] provide 60 CATS study credits equivalent.
3. Courses with the ratings: Low face-to-face teaching; Low printed materials; and High ICT-enhanced teaching [including T213 (30 CATS), T313 (30 CATS)] provide 60 CATS study credits equivalent.
4. Courses with the ratings: Low face-to-face teaching; Low printed materials; and High Online teaching [including T219 (30 CATS), T319 (30 CATS)] provide 60 CATS study credits equivalent.

The results of using the SusTEACH's Tool's Function 1 to estimate the impacts of each group of courses are presented in Table 2.

Table 2:   Average energy and carbon impacts of OU's BSc programme

Teaching Models Designs	CATS	Teaching Model impacts per student per 10 CATS		Total BSc impacts per student per 360 CATS	
		energy con-sumption (MJ)	carbon emission (kg CO_2)	energy con-sumption (MJ)	carbon emission (kg CO_2)
Low Face-to-Face, High print, Medium ICT-enhanced	180	476	42	8568	749
Low Face-to-Face, High print, High ICT-enhanced	60	536	50	3215	298
Low Face-to-Face, Low print, High ICT-enhanced	60	471	45	2829	269
Low Face-to-Face, Low print, High Online	60	451	42	2707	250
Total qualification impacts	360			17319	1566

The overall impacts for the BSc qualification programme were estimated to be 17319 MJ and 1566Kg CO_2 per student per 3600 study hours/360 CATS credits. The main sources of energy consumption and emissions for this programme were found to be attributable to campus site operations and the purchase and use of computer technologies for course activities (Figure 1).

The SusTEACH Tool may also be used to predict the effect of changing the design of a course, such as the OU's plans to replace most of the print-based delivery component of two of the BSc programme courses (S216 and T307 in Table 1) with an online learning design, blended with some minimal face-to-face teaching. In doing so, it should be noted that the main driver for change is pedagogical, and to provide individualised, flexible, interactive, synchronous and collaborative learning experiences, using computer technologies and online media, rather than primarily to reduce carbon emissions.

Within the next few years, the design of the delivery of these courses (S206 in place of S216 and T317 in place of T307) will change them from High to Low print, Medium ICT-enhanced to Online and continue to have a Low Face-to-face teaching component with minor contact delivery (a few day schools) as offered to students. Analysis using this Tool shows that this new design is estimated to reduce the carbon impacts of paper and printed materials from 8 to 3 kg CO_2 per 10 CATS credits, whilst increasing the ICT-related impacts from 8 to 13kg CO_2 per 10 CATS credits. This suggests a trade-off between the in-course ICT-related energy impacts, and the impacts of paper and printed materials.

Figure 1: Percentage sources of CO_2 in the BSc programme

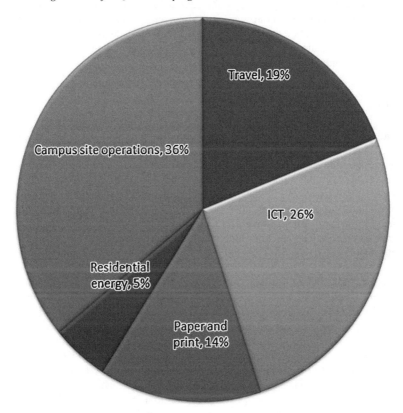

Using the Qualification Programme Teaching Models panel

Whilst the new planned course designs may not lead to carbon reductions, the overall carbon impacts associated with the various teaching models are low relative to campus-based, face-to-face teaching models. This was demonstrated by using the second function of SusTEACH Tool to compare different teaching models using the Qualification Programme Teaching

Models panel. The BSc programme presents two courses (T219 and T319) using Online Teaching Models (60 CATS credits) and the remaining six courses are taught with various ICT-enhanced Distance Teaching Models (300 CATS credits). Using this information, Sus-TEACH's Qualification Programme Teaching Models panel allowed for a ready calculation of the likely impacts using different teaching models for presenting the BSc programme.

This modelling function shows that if the full BSc qualification programme was provided using the Face-to-face Teaching Model, then the overall impacts would substantially increase in terms of energy consumption, by 89% to 154559 MJ and carbon emissions by 84% to 10005Kg CO_2 per student per 360 CATS credits. This was primarily a result of the impacts associated with the typical requirements for campus-based students to travel to university sites, establish additional residential accommodation, and use campus buildings and facilities. By comparison, a fully online qualification programme would decrease the BSc programme's estimated energy consumption by 24%, to 13069 MJ and carbon emissions, by 18%, to 1285Kg CO_2 per student per 360 CATS credits.

Conclusions

Further advice on using the SusTEACH Modelling Tool can be freely downloaded, together with the user guide, via a new online teaching unit entitled 'The environmental impact of teaching and learning' for the Open University's free Open Learn website *http://www.open. edu/openlearn/nature-environment/the-environment/the-environmental-impact-teaching-and-learning/content-section-0*. With the increased availability of computer technologies and online media, and adoption of innovative pedagogical designs in HE, the SusTEACH toolkit should have value for helping to assess how these impact on the environmental sustainability of the teaching and learning practices of the institutions offering them.

The primary value of the SusTEACH project is its novelty in determining the likely carbon-based environmental impacts of HE courses using online, ICT-enhanced and other teaching delivery models. From the analysis of the main sources of energy consumption and carbon emissions based on the SusTEACH findings in this and other studies, a Toolkit has been developed to include the SusTEACH Modelling Tool (*http://www9.open.ac.uk/susteach/background.htm#downloads*. This aims to help HE lecturers and senior managers to plan for future sustainable teaching and learning initiatives in courses and qualification programmes.

As pedagogical use of computer technologies continues to transform HE, together with future changes expected in campus building and technology energy efficiency and carbon impacts, as well as potential changes in student and staff lifestyle behaviours, there is a need for future longitudinal research to extend data collection to a larger HE sample both within and beyond the UK. The Modelling Tool is a useful tool, albeit limited in terms of the datasets available at present, and consequently open to further development and trialling.

In summary, this chapter presents a case study of using the SusTEACH Modelling Tool to estimate the likely energy impacts of a degree programme at the OU. It shows that the use of Online and ICT-enhanced Distance teaching models in this degree programme can produce carbon reductions of 84% in comparison with a wholly face-to-face taught BSc programme. Whilst the impact of using digital infrastructures and computer technologies are higher when ICT-enhanced distance and online learning designs are adopted, the overall carbon impacts of HE courses are lower as a result of reducing the main sources of energy consumption associated with teaching and learning. Moving to a fully online qualification programme would reduce carbon emissions by a further 18%, mainly by reducing the use of printed materials and travel associated with face-to-face teaching in the BSc programme.

Environmental benefits and energy savings need to be balanced against pedagogical objectives and student satisfaction. For example, a major UK survey found that student satisfaction with university programmes is associated with face-to-face contact time with academic staff (Which?/HEPI, 2013). In addition, many students enjoy receiving print-based materials, and their behaviour suggests a preference to read materials notwithstanding their coincident availability as online-only readings (Caird et al. 2013). Various explanations for this preference have been offered (see e.g. Jabr, 2013).

Overall, however, there is little doubt that the use of computer technologies and online media to provide online and blended ICT-enhanced teaching models can achieve significant carbon reductions in the delivery of educational programmes. Benefits are not simply achieved from pedagogical use of computer technologies, but as a result of new course designs that reduce the main sources of HE energy consumption associated with travel, residential energy consumption and campus site operations. ICT-enhanced, online and distance teaching methods can result in significant carbon reductions in courses, by reducing or replacing the requirements for students to travel to classrooms, live away from home, and use campus facilities. Consequently, notwithstanding the tensions noted above, addressing the challenges of achieving the transition to sustainable, low carbon HE systems could be supported by greater attention to new pedagogical designs that aim to reduce the energy impacts associated with HE teaching and learning provision.

Acknowledgements

The SusTEACH project is funded by the Joint Information Systems Committee (Jisc) under the Jisc Greening ICT Programme. We are grateful for the support of a large number of interested participants in the project from The Open University, Cranfield University, Loughborough University and the University of Oxford. For further project information see *http://www.open.ac.uk/blogs/susteach/*. To access the Toolkit tools, resources, references and further details see *http://www9.open.ac.uk/SusTeach/*.

References

Caird, S. and Lane, A. (2013). Conceptualising the role of Information and Communication Technologies in the design of higher education teaching models used in the UK. *British Journal of Educational Technology http://dx.doi.org/10.1111/bjet.12123*.

Caird, S. Lane, A. and Swithenby, E. (2013). ICTs and the design of sustainable higher education teaching models: an environmental assessment of UK courses. In: Caeiro, S.; Leal F.W.; Jabbour, C. J. C. and Azeiteiro, U. M. eds. *Sustainability assessment tools in higher education – mapping trends and good practices at universities around the world.* Springer International Publishing Switzerland, (DOI: 10.1007/978-3-319-02375-5_21) pp375-385. Retrieved from *www.springer.com/energy/energy+efficiency/book/978-3-319-02374-8.*

Conole, G. (2013). Designing for learning in an open world: Explorations in the learning sciences, instructional systems and performance technologies. New York Heidelberg Dordrecht London: Springer.

Defra (2011). *2011 Guidelines to Defra / DECC's GHG conversion factors for company reporting,* Department of Environment, Food and Rural Affairs (Defra) and Department of Energy & Climate Change (DECC), UK. Retrieved from *https://www.gov.uk/govern ment/uploads/system/uploads/attachment_data/file/69314/pb13625-emission-factor-metho dology-paper-110905.pdf.*

ECTS (2009). *European credit transfer and accumulation ECTS users' guide.* European Communities, Brussels. Retrieved from *http //ec.europa.eu/education/lifelong-learning-policy/doc/ects/ guide_en.pdf*

Jabr, F. (2013). Why the brain prefers paper. *Scientific American* 309 (5), 48-53.

Johnson, L. Adams, S. and Cummins, M. (2012). *The New Media Consortium Horizon report: 2012 higher education edition.* Austin, Texas. Retrieved from *http://www.nmc.org/ pdf/2012-horizon-report-HE.pdf.*

Lane, A. (2010). *Global trends in the development and use of open educational resources to reform educational practices.* Commissioned policy briefing for UNESCO Institute for Information Technologies in Education, 12 pp. Retrieved from *http://iite.unesco.org/policy_ briefs/.*

Roy, R. Potter, S. Yarrow, K and Smith, M. (2005). Factor 10 Visions Project: towards sustainable higher education: environmental impacts of campus-based and distance higher education systems. Final Report DIG-08, Design Innovation Group, The Open University UK. Retrieved from *http://www3.open.ac.uk/events/3/2005331_47403_o1.pdf.*

Sharples, M., McAndrew, P., Weller, M., Ferguson, R., FitzGerald, E., Hirst, T., Mor, Y., Gaved, M. and Whitelock, D. (2012). *Innovating pedagogy 2012: Exploring new forms of teaching, learning and assessment to guide educators and policy makers.* Open University Innovation Report 1. The Open University, UK. Retrieved from *http://www.open.ac.uk/ personalpages/mike.sharples/Reports/Innovating_Pedagogy_report_July_2012.pdf.*

Sharples, M., McAndrew, P., Weller, M., Ferguson, R., FitzGerald, E., Hirst, T., and Gaved, M. (2013). *Innovating pedagogy 2013. Open University Innovation Report 2.* The Open University. Retrieved from *http://www.open.ac.uk/personalpages/mike.sharples/Reports/ Innovating_Peda gogy_report_2013.pdf.*

Which?/HEPI (2013). *Higher education: The student academic experience survey* May 2013 Retrieved from *http://www.hepi.ac.uk/files/1.Higher_Educational_Report.pdf.*

European Virtual Seminar on Sustainable Development: international, multi-disciplinary learning in an online social network

Joop de Kraker[1,2] and Ron Cörvers[2]

Abstract

The European Virtual Seminar on Sustainable Development (EVS) is a web-based course that aims to foster competences for sustainable development through collaborative learning in virtual, international, multi-disciplinary student teams. An important determinant of successful team collaboration is the intensity of socio-cognitive as well as socio-emotional interactions between the team members. The extent to which a learning environment facilitates socio-emotional interactions is termed its sociability. The chapter focuses on the recent adoption of a social networking platform to enhance the sociability of the EVS virtual learning environment. The new learning environment is evaluated in terms of student experiences and perceptions, actual tool use, and team performance. The main conclusions of this evaluation are that: students hardly use the asynchronous tools for personal (non-task, socio-emotional) interaction; students have a strong preference for synchronous communication tools, which appear to support socio-cognitive as well as socio-emotional team interactions very effectively; and shifting from a traditional learning management system to a social networking platform had no measurable effect on the performance of the virtual student teams. Yet, we expect that employing a social networking platform for EVS will still be needed to match the technology expectations of upcoming generations of students, whereas finding asynchronous tools supporting socio-emotional interaction will still be required to cater for students with incompatible time schedules.

Introduction

The European Virtual Seminar on Sustainable Development (EVS) is an academic Masters-level course that can be characterized in many different ways: international, multidisciplinary, networked, and collaborative. In the context of this book on e-learning and sustainability, this chapter will focus on how EVS fosters 'learning for sustainable development', and in particular how this is supported by a custom-made e-learning environment. EVS aims at development of competences for sustainable development through collaborative learning in virtual, international, multi-disciplinary student teams (De Kraker et al., 2014), and has been widely acknowledged as 'best practice' at national (e.g., QANU, 2007; VLIR, 2007) and international level (e.g. Bijnens et al., 2006; Brey, 2007; EADTU, 2012). EVS is a unique course, being the only course in its class that has been running successfully already for over a decade (since 2001),

---

1   School of Science, Open Universiteit, Heerlen, The Netherlands.
2   ICIS, Maastricht University, Maastricht, The Netherlands.

without interruption and without external funding for its operation. Comparable courses with similar aims have been piloted, but not continued (Barth and Rieckmann, 2009; Wiek et al., 2013).

Many factors can be identified that contribute to the EVS' success, including its organisational model (Cörvers and De Kraker, 2009), and the added value it provides at student, teacher and institutional level (De Kraker and Cörvers, 2012). An additional, important success factor is that EVS has never been considered as a finished product, but has always been treated as a continuous innovation project (Cörvers et al., 2007). Over the years, a range of minor and major modifications have been implemented in the format of EVS, based on student reflection reports and teacher evaluations as well as insights obtained from research by educational scientists in EVS (e.g. Prins et al., 2005; Rusman et al., 2009; Schoonenboom, 2008; Sluijsmans et al., 2006). A recent and major change in EVS concerned shifting the virtual learning environment from a traditional learning management system to a social networking platform. This shift was motivated by the importance of social interactions for successful collaborative learning in virtual student teams (Kreijns et al., 2003; Kreijns et al., 2013a). After a description of the organisational and educational aspects of EVS, the chapter focuses on this redesign of EVS, including an evaluation based on two years of use. We end the chapter with summarizing and discussing the findings of the evaluation in terms of student experiences and perceptions, actual tool use, and team performance.

EVS: organisational model and educational format

The overall aim of EVS is to foster transboundary competence through collaborative learning in virtual, international, multi-disciplinary student teams. Transboundary competence is the ability for productive interaction across the boundaries between perspectives informed by, for example, different disciplinary or cultural backgrounds. Effective solutions to complex sustainability problems require the integration of insights from multiple perspectives, and thus transboundary competence can be considered a key competence for sustainable development (De Kraker et al., 2007, 2014). According to the principles of competence-based learning, transboundary competence can best be developed through collaborative learning in teams in a cross-boundary context. In EVS, teams in a cross-boundary context are created through virtual mobility, i.e., using a web-based learning platform to bring geographically distributed students and teachers together without the need to travel. The EVS model to organize virtual mobility is presented, followed by an outline of the EVS educational format for collaborative learning.

EVS organisational model

EVS is a Masters-level course offered by a partnership of European universities, coordinated by Open Universiteit (Heerlen, The Netherlands). The partnership is flexible, consisting of an original core group of six universities and a variable number of other universities (table 1). Since 2009, the composition of the partnership is quite stable, with currently 11 partners, representing a fruitful mix of regular and distance learning universities and an even geographic spread across Europe (table 2).

Table 1:   EVS in figures (2001-2013)

Run[1]	Participating universities	Countries represented	Students enrolled	Student teams
2001	9	4	59	6
2002	11	5	45	6
2003	15	9	61	11
2004	18	11	78	13
2005	12	9	68	10
2006	9	8	36	8
2007	6	5	22	3
2008	13	11	58	8
2009	11	9	74	8
2010	12	10	88	10
2011	12	10	73	9
2012	12	9	74	9
2013	11	9	61	9

1 Starting year: EVS runs from October to March

Table 2:   EVS partnership composition as of 2013, indicating for each partner the year of joining EVS

Year	University	Unit	City	Country	Type of university
2001	Open Universiteit	School of Science	Heerlen	The Netherlands	distance
2001	Leuphana Universität	Institute for Environmental & Sustainability Communication	Lüneburg	Germany	regular
2001	Karl Franzens Universität	Department of Geography & Regional Science	Graz	Austria	regular
2001	Charles University	Environment Center	Prague	Czech Republic	regular
2001	University of Antwerp	Institute of Environment & Sustainable Development	Antwerp	Belgium	regular
2003	University of Bucharest	Faculty of Geology & Geophysics	Bucharest	Romania	regular
2008	University of the Aegean	Department of Environment	Mytilene	Greece	regular
2008	University of Maribor	Faculty of Chemistry & Chemical Engineering	Maribor	Slovenia	regular
2008	Universidade Aberta	Department of Sciences & Technology	Lisbon	Portugal	distance
2008	Carl von Ossietzky Universität	Center for Environmental & Sustainability Research	Oldenburg	Germany	regular
2009	FernUniversität	Institut für Politikwissenschaft	Hagen	Germany	distance

The organisational model for EVS is a bottom-up network approach with distributed responsibilities, operating without formal, top-down institutional arrangements and without external financial support. EVS is a joint course, based on a the principle of collectively investing staff time. Only the initial development of EVS and the recent design and implementation of a social networking platform were partially supported by grants from the European DG Education & Culture. The distribution of tasks and responsibilities over the partners is differentiated, as institutions can become partners in an EVS run at three different levels: (1) providing students and an institutional coordinator, (2) providing one or more tutors in addition to (1), and (3) supplying a case study and providing an expert in addition to (2). The idea is that a new EVS partner starts at the first level before – hopefully after positive experiences – moving on to level two, and finally, to level three. It is up to each institution to decide at which level it wishes to start, and when it wants to switch to a different level of participation, becoming either more or less involved in the EVS. The central EVS coordinator (Open Universiteit) is responsible for the overall management and the development of EVS, as well as for maintaining the virtual learning environment. The other partners tutor and assess the student teams, develop case studies, implement EVS at their institutions (as a formally accredited, compulsory or optional course), and recruit and select students. Each participating university appoints an institutional coordinator who is responsible for the management and administration issues for his/her EVS students (intake procedure, student details, credit points etc.). Since students participating in EVS continue to be regular students of their home universities, no formal enrolment at a foreign university is necessary. Neither do they have to pay any extra fees for participating in EVS.

EVS educational format

The educational format of EVS is based on the pedagogical principles of computer-supported collaborative learning (CSCL). In CSCL, computer network technology is applied to support a type of learning characterized by:

– active learning, with students participating in small-group activities, aimed at a common goal (joint problem solving);
– students being responsible for their own learning process, with teachers taking the role of facilitators, stimulating the students to reflect on their learning;
– development of social and team skills, including consensus building (Kirschner, 2001).

The main reasons to implement CSCL are that it enables the learners and teachers to be geographically dispersed ('independence of place') and allows them to engage in the learning process at any time ('independence of time') (Kirschner and Erkens, 2013). Educational formats of CSCL-courses can be characterized by a combination of three elements (Kirschner and Erkens, 2013): the pedagogical element (referring to the role of teachers and the type of learning tasks), the technological element (referring to the computer-based tools supporting the learning process), and the social element (referring to the structuring of the collaboration process). In EVS, these elements of CSCL are translated in the 4T-model, with teachers, tasks and tools as key components, jointly supporting collaborative learning of students in teams, the 'fourth T' (figure 1).

Figure 1: The 4T-model: the four key components of the educational format of EVS

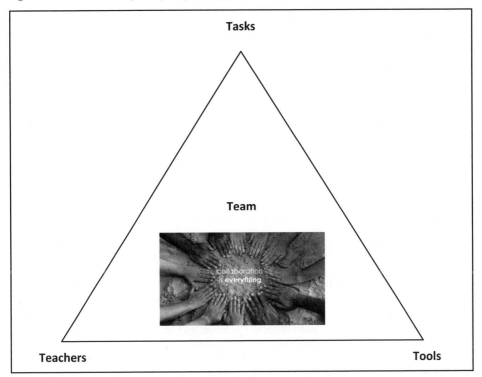

Students of different nationalities, universities and disciplines, work in virtual teams on case studies. In EVS, a case study focuses on aspects of sustainable development in one or more European countries (e.g., conservation of geo- and biodiversity, urban waste prevention, sustainable hydropower, climate adaptation, sustainable tourism). The students are provided with a brief, open-ended problem description, on the basis of which they have to develop and execute a case study research proposal. The ultimate objective is to come up with recommendations on how to contribute to sustainable development for the topic of the case study. The challenge is to make use, as much as possible, of the international and multi-disciplinary diversity in the team in terms of different perspectives on the problem and different knowledge and expertise possessed by the team members.

The total study load of EVS is approximately 120 hours, which equals three weeks of full-time study. However, the seminar is spread over a five-month period, to create better conditions for the relatively time-consuming virtual collaborative learning process and to allow students to participate in EVS alongside their regular study program or job. The core task of the students in EVS, conducting as a team a case study in sustainable development, is structured into five distinct stages, involving specific learning (sub)tasks including reflection (table 3).

Table 3: EVS learning tasks, structured into five stages

Stage	Period	Study load	Tasks	Deliverables
Registration & orientation	October-November	5 h	Registration on EVS platform; Individual activities	Personal profile page
Team formation & activities	November	10 h	Discuss SD concept; Apply to case study, Inventory of expertise	Case-specific definition of SD
Research proposal	December	25 h	Joint development of research proposal	Draft + final research proposal
Research & report writing	January-February	50 h	Literature and data collection; Analysis of results; Discussion of conclusions; Draft report writing	Draft group research report
Final report & presentation	February-March	30 h	Rewriting draft report; Making presentation; Reflection on group process	Final research report; PPT presentation; Group reflection report

The teams themselves are responsible for keeping their learning process going and delivering high-quality products by the deadlines set. To achieve this, each team is supported by a tutor and an expert. Throughout EVS, the role of the tutor is to moderate discussions, safeguard deadlines and explain what is expected from the students. It is also the task of the tutor to help solving problems in team cooperation. Alongside the tutor, each team has access to an expert, i.e., the author of the case study on which the team is working. Compared with the tutor, the expert plays a rather passive role. The main role of the expert is to give feedback on the draft versions of the research proposal and research report and to assess the final report. The tutor is responsible for assessing the team process and the team members' individual contribution to the team products. The final mark is based on a combination of these three assessments (table 4).

Table 4: Assessment criteria in EVS (contribution to final mark).

Team report (50%)
– operationalization of sustainable development
– problem analysis
– consistency of content & integration of perspectives
– application of results to target group
– editorial quality
Team process (20%)
– planning and progress, dealing with feedback
– cooperation, handling of problems
– knowledge sharing, handling of different perspectives
Individual performance (30%)
– individual planning and progress
– contribution to team products
– support to team members

Collaboration in EVS is supported by a web-based platform which offers a range of facilities and tools for communication and collaboration. The platform also provides access to all relevant course materials. In the next section, we will describe the EVS platform in more detail, and in particular focus on the recent shift from a traditional learning management system to a social networking platform.

A social networking platform for EVS

Background

Although widely implemented, CSCL has not always proven the wonder-tool that educators envisioned, and simply placing students in a group and assigning them a common task does not guarantee that they will work and learn together (Kirschner and Erkens, 2013). Also EVS has not been without problems. For example, there have been runs with drop-out rates above 40%, persistent team conflicts, and final reports lacking any form of integration of individual contributions. Kreijns et al. (2003) related such problems to a lack of social interaction in CSCL environments, and identified two pitfalls in CSCL design causing such a lack: (1) the assumption that social interaction will automatically occur because the learning environment permits it, and (2) only interactions related to the cognitive aspects of learning are stimulated, thereby neglecting the importance of socio-emotional interactions. The latter are important for the development of a sound social space, i.e., a group atmosphere characterized by trust, respect, cohesion, belonging and sense of community. A sound social space, in turn, is a precondition for effective performance of collaborative tasks. According to Kreijns et al. (2003), the pitfalls in CSCL design can be avoided by paying attention to the role of the teachers (focus on facilitation of team learning), the design of the tasks (creating positive interdependence among team members), and the tools of the learning environment (supporting interactivity). With respect to the tools, Kreijns et al. (2003) argued that CSCL environments should be made more 'sociable', as opposed to purely functional. The sociability of a CSCL environment is the extent to which it facilitates socio-emotional interactions and, through that, the emergence of a sound social space (Kreijns et al., 2013a). The specific properties of a CSCL environment that determine its sociability, are called 'social affordances'. In EVS' history of continuous evaluation and updating, there have been repeated attempts to make the virtual learning environment more sociable. These attempts focused on enhancing the 'social presence' of the student members of the virtual teams. Social presence is the extent to which the other in the communication is experienced as a 'real' person (Kreijns et al., 2013a).

One way to enhance social presence in EVS was the introduction of student profiles in the form of Word documents with personal information and a picture included. According to Rusman et al. (2009), these profiles helped students to form a mental picture of their team members and lowered the threshold to contacting each other, which contributed to the emergence of a sense of community in the start-up phase of the group work. Another approach to enhance social presence is by focusing on 'group awareness', i.e., the condition in which a group member perceives the presence of the others and where these others can be identified as discernible persons with whom communication can be initiated (Kirschner and Kreijns, 2005). An online tool, the 'group awareness widget', was developed that graphically displayed the current activities and activity patterns of the group members, which was expected to stimulate interaction, for example between members that are simultaneously online. However, the widget never made it beyond the prototype testing phase (Kirschner and Kreijns, 2005), and has not been implemented in the EVS environment.

More recently, the emergence of Web 2.0 (O'Reilly, 2005), the 'social version' of the worldwide web based on extensive application of social software, has resulted in growing attention among CSCL researchers and practitioners for new opportunities to enhance the sociability of CSCL environments (McLoughlin and Lee, 2007; Ramirez-Velarde and Alexandrov, 2007; Wheeler, 2008). Especially the spectacular rise of social networking sites, such as Facebook, with millions of users, posed the question whether social networking software could provide the solution to the lack of socio-emotional interaction in CSCL. First experiences appeared promising (Hoffman, 2009), and with EU-funding a social networking platform for EVS was developed and piloted in the context of the Lifelong Learning projects 3-LENSUS (De Kraker et al., 2011) and NetCU (De Kraker and Cörvers, 2012).

Design and implementation

During the first 10 years of EVS, Blackboard was used as the platform for the virtual learning environment (for details see Cörvers et al., 2007). Blackboard is a widely used educational content management system (or 'learning management system'), which offers a highly structured, teacher-managed course environment (*www.blackboard.com*). Although adequate for EVS, Blackboard is not very suitable for supporting social interactions. Therefore we designed a new CSCL learning environment for EVS based on social networking software (for details, see De Kraker et al., 2011). After a comparison of five alternatives, the open source software Elgg was chosen to develop and implement the social networking platform for EVS (*http://elgg.org/*).

Figure 2: Main elements of the EVS-Elgg learning platform, interconnected through information streams.

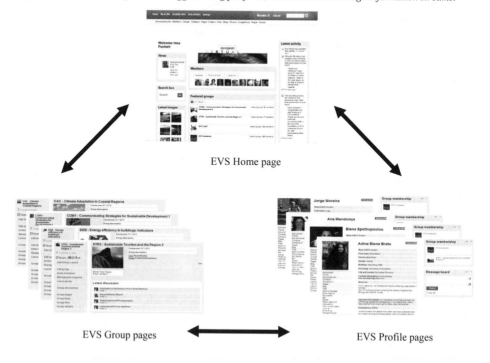

EVS Home page

EVS Group pages ⟷ EVS Profile pages

The EVS-Elgg platform is composed of three main elements: a central home page, group pages, and member pages (figure 2). The central home page is both the external, public face of the EVS community and the portal to the community. As the external face of the EVS community, it gives visitors a flavour of the community: the members, the activities, the teams and the topics discussed in EVS. For the members of the EVS community it functions as a portal. After login, it provides access to personal profiles, group pages and tools of the platform. It also gives 'inside information' about the state of the EVS community: news, latest activities, new members, groups, and tag cloud. The group pages is where it happens in EVS. The most important groups are the case study teams. On each group page, tools are available for connecting, informing, discussing, and sharing at team level. The profile pages for the members of the EVS community – students and staff – are very important, because they give faces to the community. This is meant both literally (with a photo) and by way of speaking (with personal information). Apart from sharing personal information in a fixed profile template, members can also use the profile page to post personal files, photos or blogs; in fact, anything they want to share with the EVS community. In addition to the internal tools, the EVS-Elgg tool bar also includes as standard options three external tools: Google Docs for collaborative work on documents, Doodle for making appointments for online meetings, and Skype to support these online, real-time group meetings.

Table 5:   Comparison of the interaction tools of two EVS learning platforms

Functionality[1]	EVS-Blackboard	EVS-Elgg
Communication tools	Discussion forum	Discussion forum
Collaboration tools	–	Collaborative editing
Live presentation tools	–	–
Publishing and sharing tools	File sharing	File sharing Web photo albums
Social and personal learning tools	–	Blogs Activity streams External sharing Profile pages Contacts management

1   Classification and terminology according to NetCu Toolbox:
     http://www.networkedcurricula.eu/categories

Table 5 presents a comparison of the availability of tools for interaction in EVS-Blackboard and EVS-Elgg. This comparison makes clear that the Elgg-based learning environment has many more 'social affordances' than the Blackboard version. Major differences concern the availability of information about who the other members of the virtual team are (profiles) and what they do (activities). As indicated, both types of information are expected to enhance group awareness and social presence. In EVS-Blackboard, profile information was provided in the form of Word documents, stored in a designated folder on the platform. In EVS-Elgg, each member has a profile page, which can be easily accessed by clicking on the avatar. The avatar is a thumbnail version of a member's profile picture, which is placed next to each contribution a member makes on the platform (discussion posts, uploaded files, comments etcetera).

Members are allowed and enabled to share personal items with others, e.g., by uploading pictures and documents on their profile page. Concerning information on the activities of others, this was simply not available in EVS-Blackboard. In EVS-Elgg, a time line with all activities of the other members is shown after log in, the profile page of each member gives an overview of their latest activities, the members that are online are listed at the home page, and, last but not least, notifications of contributions are sent to the e-mail addresses of the members. All this information can be tailored and restricted to information about the activities of, e.g., the members of one's virtual team only. This is done by making these persons 'contacts' (similar to 'friends' in Facebook). In addition to these affordances enhancing social presence, EVS-Elgg also has various features aimed at fostering a sense of community among the members of a virtual team.

The group pages of a team can only be accessed by its members, a 'group icon' and the profile pictures (avatars) of all team members are permanently visible on the group page, and the latest contributions (discussions, uploaded files) are shown. Affordances enhancing group awareness, social presence, and sense-of-community are expected to – indirectly – stimulate social interaction in EVS. A tool directly inviting social interaction in EVS-Elgg is the message board on each profile page, which makes it possible to respond for example to the personal information provided on the profile page. In fact, options for direct feedback to any type of contribution made by other members are ubiquitous in EVS-Elgg, in the form of comments, likes and ratings.

With respect to the main tools for communication and collaboration, there are no fundamental differences between EVS-Elgg and EVS-Blackboard. Both provide facilities for exchange (up- and downloading) of files, structured discussion boards, and links to the external tools Skype (for real-time communication) and GoogleDocs (for collaborative work on reports and presentations).

Evaluation: methods and results

Following the design guidelines for CSCL environments of Kirschner et al. (2004), the new, social network based design of the EVS learning environment was evaluated in three ways. Data from the two available EVS-Elgg runs (2011 and 2012), were analysed to determine (1) student perceptions and experiences, (2) actual use of the tools provided by the new learning environment, and (3) effect on team performance. All statistical tests were performed in Windows Excel, and prior to each test the data were checked for normality and equality of variance.

Student perceptions and experiences of EVS-Elgg

Data on student perceptions and experiences were derived from the group reflection reports each student team produced at the end of an EVS run. With nine teams per run, there were 18 reports available for analysis. These reflection reports have a fixed format with predetermined topics the student teams are asked to reflect upon. The content of all reports was checked for statements that referred to the technological tools supporting the learning process. For this evaluation, most of the relevant statements were found under the topic 'Your experiences with virtual learning and the technology (ICT-tools) you used', and some under the topic 'Learning process'. The statements are summarized below, per tool.

– *EVS-Elgg platform* (as a whole): 11 out of 18 reports commented on the platform. About half were positive (e.g., 'easy-to-use'), and the other half somewhat negative (e.g., 'navigation difficult'). The overall tendency was that the external tools (Skype, GoogleDocs) were an essential complement to the tools integrated in the platform.

- *Profiles*: only 2 out of 18 reports commented on the profiles. One stated that these were nice to get a good impression of the other team members, and the other that the profiles provided insufficient information to find out what the motivations of the other team members were.
- *Notifications*: 4 out of 18 reports commented on the notification tool (e-mail alerts of activities on the platform). All statements were positive ('nice', 'efficient', 'necessary'), and one team indicated that they would have liked to receive also notifications of comments made to contributions of others (a facility EVS-Elgg does not provide).
- *File exchange*: 8 out of 18 reports commented on the file exchange tool. Five were positive ('useful', 'very simple'), the remaining were critical, mainly about the lack of a folder structure, and one team indicated that they preferred to use Dropbox.
- *Discussion board*: 10 out of 18 reports commented on the discussion board. The tendency was predominantly positive, but balanced, stressing on the one hand it was an essential tool for reflective discussions and asynchronous communication between team members with different activity patterns and living in different time zones, and on the other hand that it lacked immediacy of feedback and resulted in slow and time-consuming communication.
- *Skype*: 18 out of 18 reports commented on the use of Skype. Fifteen teams were (very) positive about using Skype and three teams indicated that they hardly used it, because they found it very difficult to schedule synchronous meetings. Most of the teams using Skype indicate that they used the chat function, and some used the audio function. (The video function is not freely available for group sessions in Skype.) Skype is qualified as a tool that is 'essential', 'really important', 'very helpful' and 'most efficient'. It is said to be used for discussion of proposals and drafts, planning and division of tasks, coordination and decision-making. Several teams also highlight the socio-emotional value, stating that Skype sessions were important for a feeling of proximity and getting to know each other better, and for solving team conflicts.
- *GoogleDocs*: 11 out of 18 reports commented on the use of GoogleDocs. There is a remarkable difference between the 2011 and 2012 run, with three versus eight out of nine teams indicating use of GoogleDocs, respectively. All teams but one (in the 2011 run) are positive about using GoogleDocs, and half of them even very positive, with qualifications ranging from 'really nice' to 'indispensible', 'essential' and 'brilliant tool'. GoogleDocs is said to be used for collaborative work on the research proposal, research and reflection report and the case study presentation. Four teams report some (minor) problems, such as access problems and trouble with converting to Windows Office files.

Actual use of EVS-Elgg tools

The actual use of the tools provided on the EVS-Elgg platform was determined for each team as much as possible in a quantitative way. Below we explain per tool how use was quantified.

- *Profiles*: Per team the amount of information provided in the member profiles was calculated as the average number of words per profile. With GoogleAnalytics it was estimated how often member profiles were visited by other members of the team, when these visits took place and what the duration of these visits was. Team members visited each others profile page 1-5 times, mostly in the first month of the team work, and these visits lasted on average about one minute. Only very few students posted personal documents or pictures on their profile page, and only very rarely students posted messages on the message board of other students' profile pages in response to personal information on the page.

– *Contacts*: Making others your contacts makes it easier to keep track of their activities and contributions. Per team, the intensity of contact connections within the team was calculated as the average of the member values. Per member, this value ranged from 0 to 1 and was calculated by dividing the number of team members made contacts by the total number of fellow team members.

– *Files*: Per team, the total number of files exchanged in the group on the EVS-Elgg platform was counted.

– *Discussion posts*: Per team, the total number of posts on the group discussion board on the EVS-Elgg platform was counted.

– *Skype*: Per team, the total number of Skype sessions announced and/or reported in the group on the EVS-Elgg platform was counted.

– *GoogleDocs*: Per team, the stages of team work during which GoogleDocs was used was determined (research proposal; draft report; final report and presentation).

Table 6: Use of EVS-Elgg tools: average and range of nine teams per run

Tool	Run 2011	Run 2012	Difference between runs (t-test)
Profiles (words)	263 [160-372]	234 [198-248]	not significant (P=0.09)
Contacts (intensity)	0.67 [0.50-1.00]	0.55 [0.20-1.00]	not significant (P=0.12)
Files (shared)	35 [8-67]	51 [18-117]	not significant (P=0.11)
Discussion (posts)	179 [56-360]	354 [173-606]	highly significant (P=0.004)
Skype (sessions)	5.9 [1-12]	7.9 [3-14]	not significant (P=0.11)
GoogleDocs (stages)	0.8 [0-3]	2.2 [0-3]	significant (P=0.013)

The results are summarized and presented in table 6 as average values and ranges for the nine teams per run. A one-sided t-test was applied to determine whether there were significant (P<0.05) differences between these average values per run. The range values indicate that in both runs there were large differences in tool use between the teams. As a result, the runs only differed significantly with respect to the average number of discussion posts and the average number of stages during which GoogleDocs was used. Overall, communication among team members appears to have been more intensive during the 2012 run, despite that the amount of information provided by members in their profiles and the 'contact intensity' within teams was lower than in the 2011 run.

Effect of EVS-Elgg on team performance

To evaluate the effect of EVS-Elgg on team performance, the following indicators were determined for each team in the three most recent runs with EVS-Blackboard (2008-2010) and the two runs with EVS-Elgg (2011-2012):

– *drop-out rate* (0-1): number of students leaving the team before completion of EVS divided by the total number of students in the team;

- *process quality* (0-10): mark given by tutor for the quality of team communication and collaboration (see table 4);
- *product quality* (0-10): mark given by case study expert for the quality of the final report delivered by the team (see table 4);
- *within-team variation*: standard deviation of the marks given by the tutor to each team member for the quality of their individual performance (see table 4), a high variation being indicative of a low degree of team cohesion.

The five EVS runs included in this evaluation were quite similar in terms case study topics, number of teams (8-10), number of participating universities (11-13), and composition of the partnership (including tutors and case study experts). However, this does not automatically mean that the diversity of the student teams was also similar, and therefore two additional indicators were determined for each run, representing team diversity at the start and the end of each run of EVS. Team diversity was calculated as the number of nationalities represented in a team, divided by the total number of team members.

Table 7:   Team diversity and team performance: average values of 8-10 teams per run

Run	Drop-out rate	Process quality	Product quality	Within-team variation	Team diversity start	end
2008 Blackboard	0.43	7.3	7.8	1.3	0.75	0.76
2009 Blackboard	0.20	8.3	7.9	1.3	0.67	0.71
2010 Blackboard	0.34	8.2	7.6	1.2	0.76	0.82
2011 Elgg	0.39	8.1	7.5	1.3	0.79	0.80
2012 Elgg	0.26	8.6	8.7	1.0	0.68	0.71

The average values per run of the indicators of team diversity and team performance are presented in table 7. There are no clear differences between the three EVS-Blackboard runs on the one hand and the two EVS-Elgg runs on the other, hence no effect of type of learning environment on team performance could be established. A one-sided t-test applied to the two EVS-Elgg runs, however, revealed that there were (marginally) significant differences between these runs where it concerned 'team diversity at start' ($P=0.058$), 'drop-out rate' ($P=0.03$), and 'product quality' ($P=0.03$), indicating that higher team diversity at start was associated with a higher drop-out rate and a lower product quality. Correlation analysis across all five runs of the indicators listed in table 7, confirmed a positive correlation between the average team diversity at start and the average drop-out rate. However, when a correlation analysis was performed on these indicators at team level (all teams pooled across the runs), only a significant positive correlation between team process quality and team product quality emerged. These correlations are no proof of causal relationships, but it is safe to say that at EVS team level the quality of the team process is a good predictor of the quality of the team product, and that at EVS run level, the average team diversity at the start is a good predictor of the average drop-out rate. The fact that the latter relationship was not found at team level, indicates that other factors that differ between teams (e.g., characteristics of the students other than nationality) overrule the effect of nationality-based team diversity on the drop-out rate.

Tool use and team performance in EVS-Elgg

A final step in the evaluation of the EVS-Elgg learning environment, was to assess possible relationships between actual tool use and team performance (as presented in the previous sections). As indicated, there were several significant differences between the two runs in both respects. Concerning tool use, the average number of discussion posts and the average number of stages during which GoogleDocs was used, was significantly higher during the 2012 run, and these differences are indicative of an overall higher communication intensity in the 2012 run. In this same run, the drop-out rate was significantly lower and the product quality significantly higher than in the 2011 run. However, correlation analysis at team level (across both runs) between indicators of tool use (table 6) and indicators of team performance (table 7), only revealed significant positive correlations between the intensity of Skype use and team process and product quality. This indicates that at team level other factors that differ between teams (e.g., characteristics of the tutors, case study experts and students) have more effect on team performance than the use of specific tools other than Skype. Moreover, closer inspection of the differences in tool use between teams suggests that teams found different, alternative ways to come to a good result, e.g., some high-performing teams worked intensively in GoogleDocs, whereas other high-performing teams exchanged many files and posts on the group discussion board.

During both runs with EVS-Elgg, a psychometric questionnaire was administered at the start and the end of the run to measure the quality of the 'social space' in the virtual teams (for details see Kreijns et al., 2013b). The questionnaire was based on the validated approach by Kreijns et al. (2004), but, unfortunately, tests of the results have shown that further validation and refinement is needed (Kreijns et al., 2013b). Statements in the group reflection reports give an idea of the quality of the social space in the teams, however. Teams with the highest perceived quality of social space, as reflected in statements such as 'it was great to work together', 'we felt mutually responsible', 'the collaboration was terrific', 'we grew together quite close', and 'everyone was willing to invest', were also the teams with the highest frequency of Skype meetings. Often these teams made this connection themselves in their reflection reports. On the other hand, teams that were not able to organize regular skype meetings, indicated that the atmosphere in the team was characterized by 'continuous problems', or 'strictly professional', and that some members missed the establishment of 'virtual friendships'. Also these teams suggested themselves that there was a relationship between the two.

Conclusions and discussion

EVS aims to foster competences for sustainable development through collaborative learning in virtual, international, multi-disciplinary student teams. An important determinant of successful team collaboration is the intensity of socio-cognitive as well as socio-emotional interactions between the team members. The extent to which a learning environment facilitates socio-emotional interactions is termed its sociability. The chapter focuses on the recent adoption of a social networking platform to enhance the sociability of the EVS virtual learning environment. In this final section, we start with summarizing the findings of an evaluation of the new learning environment (EVS-Elgg) in terms of student experiences and perceptions, actual tool use, and team performance. Next, conclusions are formulated on the basis of these findings. We end this section with a discussion of the conclusions and an outlook on further development of the EVS learning environment.

Summary of findings

Students acknowledged the important function fulfilled by the asynchronic communication tools provided by the EVS-Elgg platform. However, most expressed a strong preference for synchronic, real-time communication as offered by Skype, a tool external to the EVS-Elgg platform. According to the students, the immediate feedback enabled by Skype group chat or telephone conferencing is important for fruitful socio-cognitive interactions (conclusive group discussion, task coordination) as well as socio-emotional interactions (getting to know each other better, solving conflicts). Teams with a self-reported sound social space (group atmosphere) related this to their high frequency of Skype meetings, whereas teams with a self-reported low quality of social space related this to their inability to schedule more than a few Skype meetings.

The tools offered by the EVS-Elgg platform for personal, non-task interaction between team members were hardly used. The interaction tools of the platform that were used most (exchange of files in the group environment and the group discussion board), were employed for task accomplishment. Of the external tools, Skype was used by all teams, with an average frequency of 6-8 group meetings per run. GoogleDocs was used for collaborative work, in particular in the 2012 run. In both runs with the EVS-Elgg platform, the intensity of use of the different varied greatly between teams, but on average communication and collaboration among team members were more intensive in the 2012 run as compared to the 2011 run. During the 2012 run, both team diversity at the start and the drop-out rate were lower, whereas the product quality was higher. At team level, only the intensity of Skype use was positively correlated with team process quality and team product quality, across the two EVS-Elgg runs.

No notable differences were observed in average team performance between the three most recent runs with the EVS-Blackboard platform and the two runs with the EVS-Elgg platform. Across all runs, team diversity and drop-out rate were positively correlated, but at team level, only a positive correlation between process and product quality was found.

Conclusions

On the basis of the findings of the evaluation of the EVS-Elgg learning environment, we conclude that:

– The EVS-Elgg learning environment offers extensive tools for personal (socio-emotional, non-task) interaction between students, but students hardly used these. Instead, students preferred to use the external tool Skype, which – according to them – served well not only for socio-cognitive interactions, but also fulfilled socio-emotional needs. Both student reflection reports and statistical analysis of tool use and team performance indicate that there exists a close and positive relationship between the frequency of Skype team meetings, the quality of the social space of a team, and team performance.

– Adoption of EVS-Elgg as new learning environment had no significant effect on team performance (drop-out rate, process quality, product quality) as compared to EVS-Blackboard. Across EVS-Elgg as well as EVS-Blackboard runs, poor team performance in terms of drop-out rate was related to high team diversity in terms of nationalities. However, it appears that at team level other factors (e.g., the quality of the tutor) overrule the potential negative impact of high team diversity on the drop-out rate.

Discussion

In the CSCL literature, empirical studies on the application of social networking software to support collaborative learning are scarce and difficult to compare. In contrast to our findings, Hoffman (2009) noted that many students made spontaneous use of the tools for personal interaction offered by a social networking application (Ning) that was linked to a traditional learning management system. Students shared personal images and posted comments on profile pages (Hoffman, 2009). However, Skype was not included as a synchronous communication tool in this case, which could explain the difference with our findings. The experiences in another case (Lu and Churchill, 2012), in which real-time group chat was included as a tool in an Elgg-based learning environment, do correspond with our findings. Similar to our case, Lu and Churchill (2012) observed a strong preference for synchronous communication tools, i.e., chat, for instant sharing and discussing of digital resources and efficient, quick group decision-making, whereas the Elgg-based platform was used for exchanging comments, storing of shared resources and publishing of completed tasks. Our study shows that a synchronous communication tool like Skype can support socio-cognitive as well as socio-emotional interactions in teams very effectively, so it probably should come as no surprise that students have such a strong preference for this type of tools over the asynchronous social networking tools. Moreover, in the case of high-performing teams that made intensive use of Skype, there appeared to be positive feedback relationships between team performance, quality of social space and Skype use. The members of teams that were able to plan regular Skype meetings at a relatively early stage, soon became more close, organized, effective, and willing to help each other and live up to expectations, which in turn resulted in greater willingness to invest in frequent Skype meetings. For teams that started with difficulties in the planning of synchronous group meetings, this feedback worked the other way around. This phenomenon confirms the theoretical framework of the social aspects of CSCL environments by Kreijns et al. (2013a), which includes reinforcing relationships between the intensity of social interaction on the one hand, and the quality of social space and learning outcomes on the other.

Shifting the EVS learning environment from a traditional learning management system (Blackboard) to a social networking platform (Elgg), had no measurable significant effect on the performance of virtual student teams. This observation corresponds with Russell's (1999) 'no significant difference phenomenon', stating that 'technology as such' has no significant effect on the effectiveness of learning environments. However, according to Russell (1999), technology may have other advantages that might indirectly influence the effectiveness of the learning environment. Hoffman (2009) reports such an effect of linking a social networking application (Ning) to a traditional learning management system: there was no measurable effect on the quality of the learning outcomes, but learning experience (feeling connected) and attitude (motivation) improved. Similarly in EVS-Elgg, the affordances enhancing group awareness, social presence and sense-of-community may have been effective. However, no discernible effect on the drop-out rate or the quality of the team process was found. On the other hand, we also did not find a negative effect of adoption of a social networking platform on the quality of the team reports, as anticipated by Zhang (2009) and to some extent observed by Lu and Chuchill (2012). According to Zhang (2009), the social networking approach to sharing knowledge and ideas by posting, commenting and chatting, results in an embedded and dispersed representation of team knowledge, which makes it difficult to deepen and advance the knowledge of the team. Lu and Churchill (2012) did indeed observe that the adoption of a social networking environment resulted in fragmentation of knowledge across the platform, and concluded that the environment seemed not appropriate for the construction of new knowledge. In the design of EVS-Elgg these effects were counteracted by including a structured group discussion board and by adopting applications for collaborative work on joint documents, such as GoogleDocs. This appears to have been effective.

Whereas the adoption of the EVS-Elgg platform had no impact on team performance, a remarkable outcome of the evaluation was that team diversity in terms of nationalities appeared to enhance drop-out rates and negatively affect team process and product quality. At the level of the individual teams, this effect was overruled by the variation in other factors, for example team tutor quality, but between runs, when similar collections of teams can be compared, the effect was significant. This seems to undermine the whole rationale of EVS, in which team diversity in terms of nationalities and disciplines plays a central role in the development of transboundary competence and successful production of integrated knowledge for sustainable development (De Kraker et al., 2014). However, a closer look at the data reveals that the underlying mechanism has little to do with a negative effect of diversity of nationalities as such. In practice, certain characteristics of EVS students that may affect team performance happen to be associated with their nationality. For example, the decrease in average team diversity from the 2011 to the 2012 run was mainly due to the withdrawal of the Belarussian International Sakharov Environmental University (ISEU) from the EVS partnership. The nine students from ISEU that enrolled in the 2011 run were not receiving credits, often had fulltime jobs and their internet access was of poor quality. Not surprisingly, their participation often negatively affected team performance and two-thirds of them dropped out before the end of EVS. The implication for EVS is that diversity in terms of nationalities should not be reduced, but that strict implementation is needed of the entry requirements for students (e.g., concerning time availability and internet access, see Cörvers et al., 2007).

Outlook

What do the findings mean for further improvement of the EVS learning environment? Should the social networking platform be abandoned as it has no measurable effect on team performance and Skype can effectively provide in the need for socio-emotional interactions? We don't think so. Although EVS-Elgg and EVS-Blackboard may in practice not differ significantly in pure functionality, the 'look and feel' of EVS-Elgg matches much better with the expectations the current 'Facebook generation' of students have for a web-based communication environment. We expect that not adapting to these expectations will on the longer term have negative impacts on team performance due to reduced motivation. Another reason is that asynchronous tools for social interaction remain important because not all teams are capable to plan regular synchronous meetings with Skype. Further research in subsequent runs of EVS is needed to determine the underlying causes, but in a number of cases this is simply due to incompatible day or week schedules or because team members live in very different time zones. Similar problems would occur when the EVS model would be applied by partnerships that included universities from the Americas and/or Asia. Thus, a major challenge in further development of the EVS learning environment is to find effective, asynchronous tools that support socio-emotional interaction in virtual teams when synchronous tools are not an option.

Acknowledgements

Thanks are due to our partners in EVS, 3LENSUS and NetCu. The financial support of the European Commission to the 3-LENSUS and NetCu projects is acknowledged. This article reflects the views only of the authors, and the Commission cannot be held responsible for any use which may be made of the information contained therein.

References

Barth, M., & Rieckmann, M. (2009). Experiencing the global dimension of sustainability: student dialogue in a European-Latin American virtual seminar. *International Journal of Development Education and Global Learning*, 1(3), 22-38.

Bijnens, H., Boussemaere, M., Rajagopal, K., Op de Beeck, I., & Van Petegem, W. (2006). *European cooperation in education through Virtual Mobility – a best-practice manual*. Heverlee: EUROPACE izvw.

Brey, C. (Ed.). (2007). Guide to Virtual Mobility. Report e-move project, European Association of Distance Teaching Universities (EADTU).

Cörvers, R., Leinders, J., & van Dam-Mieras, R. (2007). Virtual seminars – or how to foster an international, multidisciplinary dialogue on sustainable development. In J. de Kraker, A. Lansu & R. van Dam-Mieras (Eds.), *Crossing boundaries – Innovative learning for sustainable development in higher education* (pp. 142-187). Frankfurt am Main: VAS Verlag für Akademische Schriften.

Cörvers, R., J. de Kraker (2009). Virtual campus development on the basis of subsidiarity: The EVS approach. In: M. Stansfield and T. Connolly (Eds.), *Institutional Transformation through Best Practices in Virtual Campus Development: Advancing E-Learning Policies*, pp. 179-197, Information Science Reference, Hershey/New York: IGI Global.

de Kraker, J., Lansu, A., & van Dam-Mieras, R. (2007). Competences and competence-based learning for sustainable development. In J. de Kraker, A. Lansu & R. van Dam-Mieras (Eds.), *Crossing boundaries – Innovative learning for sustainable development in higher education* (pp. 103-114). Frankfurt am Main: VAS Verlag für Akademische Schriften.

de Kraker, J., Cörvers, R. (2012). EVS – European Virtual Seminar on Sustainable Development. In: EADTU, *NetCu Compendium of Showcases*, pp. 55-66.

de Kraker, J., Cörvers, R., Valkering, P., Hermans, M., Ruelle, C. (2011). Potential of social software to support learning networks for sustainable development. In: Barton, A., Dlouha, J. (Eds.), *Multi-actor Learning for Sustainable Development in Europe: a Handbook of Best Practice*. Grosvenor House Publishing, Guildford, pp. 124-143.

de Kraker, J., Cörvers, R., Lansu, A. (2014). E-learning for sustainable development: linking virtual mobility and transboundary competence. In: Azeiteiro, U.M., Leal Filho, W., Caeiro, S., (Eds.). *E-learning and sustainability*, Peter Lang.

EADTU (2012). NetCu Compendium of Showcases: showcases of networked curricula. Heerlen: European Association of Distance Teaching Universities.

Hoffman, E. (2009). Evaluating social networking tools for distance learning. In *TCC-Teaching Colleges and Community Worldwide Online Conference*, Vol. 2009, No. 1, pp. 92-100.

Kreijns, K., Kirschner, P. A., & Jochems, W. (2003). Identifying the pitfalls for social interaction in computer-supported collaborative learning environments: a review of the research. *Computers in Human Behavior*, 19(3), 335-353.

Kreijns, K., Kirschner, P. A., Jochems, W., & Van Buuren, H. (2004). Measuring perceived quality of social space in distributed learning groups. *Computers in Human Behavior*, 20(5), 607-632.

Kreijns, K., Kirschner, P. A., & Vermeulen, M. (2013a). Social aspects of CSCL environments: A research framework. *Educational Psychologist*, 48(4), 229-242.

Kreijns, K., Van Acker, F., Kirschner, P. A., Vermeulen, M., & van Buuren, H. (2013b). Exploring the assessment of social presence and social space in a Community of Inquiry. Paper presented at the 15th biennial European Association for Research on Learning and Instruction (EARLI) conference, August 27-31, Munich, Germany.

Kirschner, P. A. (2001). Using integrated electronic environments for collaborative teaching/ learning. *Learning and Instruction*, 10, 1-9.

Kirschner, P. A., & Kreijns, K. (2005). Enhancing sociability of computer-supported collaborative learning environments. In *Barriers and biases in computer-mediated knowledge communication* (pp. 169-191). Springer US.

Kirschner, P. A., & Erkens, G. (2013). Toward a framework for CSCL research. *Educational Psychologist*, 48(1), 1-8.

Kirschner, P., Strijbos, J. W., Kreijns, K., & Beers, P. J. (2004). Designing electronic collaborative learning environments. *Educational Technology Research and Development*, 52(3), 47-66.

Lu, J., & Churchill, D. (2012). The effect of social interaction on learning engagement in a social networking environment. *Interactive Learning Environments*, doi:10494820.2012. 680966

McLoughlin, C., & Lee, M. J. (2007). Social software and participatory learning: Pedagogical choices with technology affordances in the Web 2.0 era. In *ICT: Providing choices for learners and learning*. Proceedings ascilite Singapore 2007, pp. 664-675.

O'Reilly, T. (2005). What Is Web 2.0? Design patterns and business models for the next generation of software. Available at: *http://oreilly.com/web2/archive/what-is-web-20.html*.

Prins, F. J., Sluijsmans, D. M. A., Kirschner, P. A., & Strijbos, J.W. (2005). Formative peer assessment in a CSCL environment: A case study. *Assessment and Evaluation in Higher Education*, 30, 417-444.

QANU – Quality Assurance Netherlands Universities (2007). Onderwijsvisitatie Milieuwetenschappen. Utrecht: QANU.

Ramirez-Velarde, R. V., & Alexandrov, V. (2007). Web 2.0 Technologies applied to collaborative learning. Paper presented at ICL2007 Conference, September 26-28, 2007, Villach, Austria.

Rusman, E., van Bruggen, J., Cörvers, R., Koper, R., & Sloep, P. (2009). From pattern to practice: evaluation of a design pattern fostering trust in virtual teams. *Computers in Human Behavior*, 25(5), 1010-1019.

Russell, T.L. (1999). No Significant Difference Phenomenon (NSDP). North Carolina State University, Raleigh, NC, USA.

Schoonenboom, J. (2008). The effect of a script and a structured interface in grounding discussions. *International Journal of Computer Supported Collaborative Learning*, 3(3), 327-341.

Sluijsmans, D. M. A., Prins, F. J., & Martens, R. L. (2006). The design of competency-based performance assessment in e-learning. *Learning Environments Research*, 9(1), 45-66.

VLIR – Vlaamse Interuniversitaire Raad (2007). De onderwijsvisitatie Milieuwetenschappen. Een evaluatie van de kwaliteit van de masteropleidingen Milieuwetenschappen aan de Vlaamse universiteiten. Brussel: VLIR.

Wheeler, S. (2008). All Changing: The Social Web and the Future of Higher Education (a tale of two keynotes). Keynote speech for the Virtual University: Models, Tools and Practice Conference, June 18-20, 2008, Warsaw, Poland.

Wiek, A., Bernstein, M. J., Laubichler, M., Caniglia, G., Minteer, B., & Lang, D. J. (2013). A Global Classroom for International Sustainability Education. *Creative Education*, 4(4A), 19-28.

Zhang, J. (2009). Comments on Greenhow, Robelia, and Hughes: toward a creative social web for learners and teachers. *Educational Researcher*, 38(4), 274-279.

Electronic logistics for a sustainable distance education: the new UNED on-site virtualization of evaluation procedure documents

**Mª Carmen Ortega-Navas, Rocío Muñoz-Mansilla,
Fernando Latorre and Rosa María Martín-Aranda**[1]

Abstract

Distance education has very much changed in the last three decades. It provides instruction means for distance learning based on specific didactic methodologies. This paper illustrates how the Spanish National Distance Education University (UNED) have implemented a new protocol for evaluation procedures that optimizes paper use and transportation through on-site digitalization of exams, bringing a new logistics paradigm. The fundamental way of control of the academic performance of the students stands on the evaluation procedures: exams. UNED evaluation system simultaneously summons students at many locations in Spain and at selected venues across the World. The exams are individually delivered and then scanned via encrypted virtual system on a global scale. This work elaborates on UNED student evaluation and virtualizing system to maximize the benefits associated with electronic data handling and logistic challenges. The communication between professors and students is always open to check the evaluation of the exams. Technology is the main contribution in the so-called "valija virtual" (virtual attaché case) system, which has been developed at UNED.

Introduction

Distance education has experienced a spectacular growth since the early 1980s. It has evolved from early correspondence education using primarily printed materials into a global activity standing on multiple technologies. For example in 1971, Open University- UK which used a mix of audio-visual print. There is not *a single* concept for distance education, but a variety of such concepts (table 1). Distance education for teaching/ learning processes has been revised by different authors and researchers over the years (Litwin, 2000; Barbera et al., 2001; Talnot, 2006; Hope and Guiton, 2006; Bernath, 2009; Cleveland-Innes, 2010; García Aretio, 2011; Moore and Kearsley, 2012; Moreno, 2012). However, there are some basic common features: the separation of professor and learner which distinguishes it from face to face lecturing; the use of technical media; and the provision of two-way communication in ways that facilitate learning in distance.

---

1   National Distance Education University (UNED), c/ Bravo Murillo, 38, 28015, Madrid (SPAIN).

Table 1:   Summary of distance education definitions

Authors	Definitions
Dan (2000)	Education method in which the learner is physically separated from the professor and the teaching institution
Casarotti, Filliponi, Pieti and Sartori (2002)	The professor and the students are separate in the spatial dimension and that this distance gap is filled by technological means
Simonson (2003)	Institution-based, formal education where the learning group is separated from the teaching group and where interactive telecommunications systems are used to connect learners, resources, and instructors
Schlosser & Simonson (2006)	An institution-based formal education where the learning group is separated, and where interactive telecommunications systems are used to connect learners, resources, and instructors.
Moore and Kearsley (2012)	Teaching and planned learning in which teaching normally occurs in a different place from learning, requiring communication through technologies as well as a special institutional organization.

Some of the most common characteristics of distance education are:

- Distance education is a two-way communication system to flexibly reach a large number of students scattered in space and time. This requires new roles for students and professors, new attitudes and new methodological approaches.
- Distance education modifies the academic formal models of conventional education; it seeks for students' autonomous learning adapting to their needs and schedules. It takes advantage of Information Technologies and Communication tools to improve their independent learning and facilitate communication with their tutors.

This chapter illustrates the Spanish National Distance Education University (UNED) latest methodology for evaluation procedures. UNED was founded in 1972 as a higher education center accessible to everyone. UNED is one of the largest European Universities with around 260,000 students, and more than 1,400 professors; it also has nearly 7,000 professor-tutors spread across the Associated Centers in different towns. UNED has a wide important social support function and it has established a wide network with sixty-one associated centers in Spain and seventeen worldwide. In addition, it delivers many so-called "open courses" under the "UNED Abierta" umbrella, which do not belong to a given official degree. "UNED Abierta" initiative is based on organizing and facilitating access to an extensive range of non-formal educational materials and lifelong learning continuous education.

UNED integrates the new Information and Communication Technologies (ICT) Society. These are implemented for synchronous and asynchronous uses; such as virtual learning environments that support both the student-professor interaction and the collaborative processes among students (WebCT, aLF); university internet portals; interactive learning programs (e.g., language learning), videoconferences, or educational TV programs. In this chapter, we focus on the new evaluation protocols, which take advantage of *on-site digitalization of exams*. This is called *"valija virtual" (virtual attaché case)* in coordination with the Universal Mobile Telecommunications System (UMTS) and Wireless Application Protocol (WAP) technologies.

The *virtual attaché* procedures optimize paper use and transportation costs through on-site digitalization of exams, bringing a new logistics paradigm. The *virtual attaché* system is an UNED an innovative project delivering greater security and safety to the large number of tests performed at the UNED (ca. 780,000 exams in 2013). The great value is that exams are directly sent to evaluators in a digital manner. This speeds up evaluation procedures for our Bachelor and Masters degrees. The original documents remain thus safely stored at the associated centers at the venue of the exams.

The UNED *virtual attaché* contributes to the sustainability in teaching, implementing the UNED 2010 sustainability program, which focuses in an optimization of energetic, time and material resources.

Information and Communication Technologies (ICT) at UNED University

UNED distance education combines the use of print and audiovisual media with new technologies. 264,059 students enrolled our university in the 2012-2013 academic year. There is a high number of foreign students, 7,518, belonging from 127 countries. UNED geographical outreach is illustrated by near 3,100 students distributed among the seventeen centers worldwide (in Bata, Bern, Brussels, Buenos Aires, Caracas, Berlin, Lima, London, Malabo, Mexico City, Paris, São Paolo, New York, Rome, Bogota, and Santiago de Chile).

On the other hand, UNED university servers support 5,213 online courses in the virtual learning program delivering over 27 degrees, 54 masters and other non-formal education courses.

Students can also communicate with their professor via AVIP (IP virtual class) classrooms. AVIP is a synchronous teaching tool that provides technological support for all educational activities. AVIP enables "virtual presence" giving a next to real presence education experience (Figure 1). AVIP classrooms offer more than 37,000 videos and 33,000 on-line conference recordings that exceed 5 ½ million visits.

YouTube channel and similar services are a particularly valuable means for remote presence in the network. These have been viewed near 5,000,000 times, with more than 12,000 followers during the course 2012-2013. This increase is mainly due to the COMA (Mass and Open Online Courses) courses, which fall under the "Open UNED" program, which is committed for repositories dissemination. There have been over 204,000 enrollments and COMA UNED platform surpassed 1,600,000 visits worldwide in September'13. The UNED is also present in ITunesU, with 24 collections and 9 courses, which can be easily downloaded and played on any Apple or PC devices.

Figure 1: AVIP-based professor/student virtual presence at distance seminars and tutorials in a remote associated center, UNED.

Evolution during the time and need for a more sustainable and economic approach

Sustainability becomes increasingly important, due to its impact on the environment and the society. University sustainability assessment frameworks are gaining "popularity with an increasing number of universities demonstrating leadership on combating climate change and pursuing sustainability" (Shi and Lai, 2013, p. 59). Sustainability processes take time, which implementation in education for sustainable development (ESD) is "a tough challenge and actual results are still far from the desired image of a higher ESD" (Holmberg et al., 2012, p. 220).

Universities are increasingly promoting sustainability instruments (Caeiro et al., 2013). These are sustainability assessment tools specifically developed for higher education institutions, such as the *Auditing Instrument for Sustainability in Higher Education* (AISHE), the *Graphical Assessment of Sustainability in Universities* (GASU) or the *Sustainability tool for Auditing Universities Curricula in Higher Education* (STAUNCH).

The number of exams at UNED has continuously increased during the last years: our 260,000 students have done more than 781,000 exams during the 2012-13 academic year (Figure 2). Such a dimension demands a paradigm shift towards more sustainable evaluation procedures.

Figure 2: Number of exams per year from 2005 to 2013

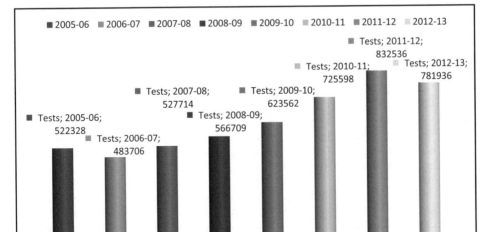

Monitoring these exams requires a major economic, material and human resources effort. Exams are made at the UNED Associated Center venues across Spain and worldwide. Evaluation panels constituted by professors from the University Faculties in Madrid go those venues on selected dates: two weeks in February, two weeks in June and one week in September. In an earlier stage, we developed the use of the "*Valija Virtual*" (Virtual Attaché) for Spain in 2007-08 academic year. That was to implement an efficient distribution of exams based on electronic recording of these and on site production of physical exams at the examination venue. This would prevent the energy intensive transportation of exam hard copies from Madrid headquarters to the corresponding venues. This system was then implemented at the worldwide venues; first at the Brussels Associate Center in the 2009-10 course; then, the *virtual attaché* has been implemented in London, Paris and Berlin during the 2013-14 academic course.

The next step was to be able not only to electronically deliver the exams but to electronically collect them. There are multiple benefits. Exams can readily be at a secure server, where professors and students may access, and transportation costs are minimized since the original exams remain guarded at the venue where these have been produced.

New protocol for evaluation procedures: *Virtual Attaché* system

The *Virtual Attaché* ("Valija Virtual") application has been totally developed at our Barbastro Computer Department Center. The objective is to simplify the processes of preparing, transporting, distributing and collecting exams, maximizing security guarantees. This computer system is used in all UNED centers to conduct exams. The entire test management process is digitally controlled delivering major benefits, such as:

– Miniamized costs and paper use;
– Improved time management;

- Greater security;
- Supporting the evaluation panel work;
- Increased student satisfaction.

Feature

UNED students do their exams at the associated center venues geographically distributed across Spain and worldwide. The exams occur simultaneously everywhere and students go to the venue that is most convenient to them. On arrival, students are identified by their corresponding barcode UNED ID document (Figure 3a); alternatively they may insert their electronic national ID (Figure 3b). The system automatically prints out the corresponding exam pages. The system also distributes students taking the same test apart from each other at the exam room.

Students hand their completed exam to the evaluation panel members at the end of the exam. These exams are then scanned on site, creating a backup copy. Such scanned files are sent directly to each student and to the corresponding professor after two days. Upon authentication, students can access and view all their answers given on the exams, from home. Thus, professors have readily access to all exams via a secured access. We have developed an evaluation program that allows the automatic correction of multiple-choice exams. Such a virtual attaché thus delivers not only efficient and safe timely delivery of exams to professors, but it also provides a sustainable value on paper use and transportation costs.

Implementation

We made a stepwise implementation. Starting with pilot testing. The 2010 May-June exam call combined the experimental electronic exam scan along with conventional exam procedures.

The *virtual attaché* was introduced in all centers in successive calls during the 2010-2011 course. This included the digitization (scanning) of all tests. This has been validated throughout the 2011-12 course. Due to its population, Madrid is one of the largest UNED evaluation venues. Three of the five centers of exams in Madrid successfully joined the digital return of exams in September 2013. More than 500 exams were rapidly and safely processed at the same time in each of these venues. Finally, it got generally implemented for all evaluation panels in continental and insular Spain during the 2012-13 academic year.

A built-in security code for each exam was implemented in 2012 to increase the guarantees; it prevents the evaluation panels from wasting time marking individual exams. This is particularly important if we consider the large number of tests done across the country and worldwide by UNED.

Tests hardcopies no longer return to UNED headquarters in Madrid. These remain at each local UNED Associated Center venue where the exams were made, and are guarded during at least two years before their destruction.

Procedure

Exams at the UNED Associated Center venues involve collaboration between the local team and the university team that goes there to execute the evaluation. The *Virtual Attaché* digitally controls the entire test management process, using two user profiles: local management staff and academic members of the evaluation panel; this enables full management of the development of each evaluation session.

Figure 3: a) University Card of employers of UNED, containing electronic signature

Figure 3: b) National identity card of Spain (dni-e), with electronic signature

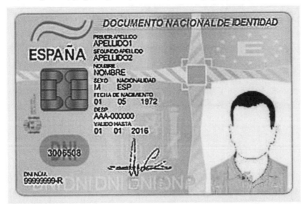

Students are identified by either photo ID: their university ID (Figure 3a) or the Spanish electronic national ID (Figure 3b). Both cards contain the pertinent electronic certificates. In particular, the Spanish electronic national ID is issued by the Spanish Public Authority of Certification (CERES) reporting to the FNMT-RCM (Spanish Mint) under the terms of Act 59/2003, of 19 December, on Electronic Signatures. The procedural steps are as follows:

Prior to the exam session

Encrypted examinations can be decrypted in two ways. In an earlier stage, examinations were stored in a CD. The unlocking key disc was guarded by the Evaluation Panel chair. In a later stage, decryption is progressively enabled during 2011-2012 year by an electronic certificate; this is in the professor's smart card. The latter will progressively replace the former, currently in use. Thus, greater security and reduction of CD usage will be increasingly in force.

Access to and location at the exam venue

The application allows selecting the session date and time so that the relevant students have access to their exam class. A reader reads the card of each student recognizing the exams that he/she is attending. A fast printer prints both examination questions and pages to be used for answering. An intuitive traffic-light based indicator tells when the next student may enter (Figure 4). For a more efficient operation, this application is compatible with several simultaneous entry controls at the exam room.

Figure 4: Representative view of the access control application window for student access.

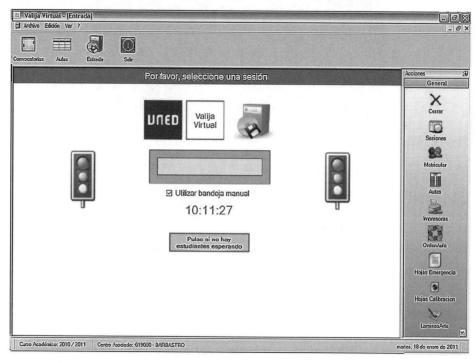

The printed test header not only identifies the student and tells the specific exam but it also indicates the student the time available for its completion and, if applicable, the materials that can be used during the test (Figure 5). It also indicates the position to be occupied at the exam room so that students doing the same exam do not sit adjacent to each other.

Figure 5: Sample of printed exam header.

1111 Nombre Apellido 1 Apellido 2	Junio 2008	00 - ACCESO		
		001 - LENGUA ESPAÑOLA		
		18/05/2008 Hora de entrada: 09:00 Hora de salida: 13:51	Examen tipo: B Lectura óptica	Aula Indefinida Fila: 0 Columna:0
BARBASTRO - 019000			NACIONAL PRIMER TURNO	Hoja 1 de 3
Material:/h~lä Ninguno				

Es imprescindible entregar esta hoja para salir del aula.
NO ESCRIBIR EN EL REVERSO DE ESTA HOJA   ¿Desea obtener un certificado de asistencia? ☐

During the exam session

The application permits the evaluation panel to control the development of the exam session for it allows (Figure 6):

- identifying the student who occupies each position in the exam room;
- individual control of time available for each student, with a system of notices when a student has exceeded his/her time;
- knowing the students remaining to hand back a given exam;
- getting listings of the students presented.

Figure 6: Representative individual control of time available to each student and system of notices

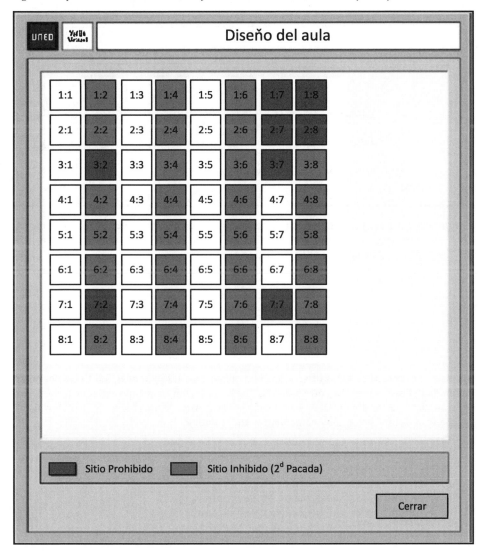

Virtual Attaché Return

It is a computerized system that complements the Virtual Attaché. It is designed to facilitate the professors' access to the examinations done by their students. This minimizes transportation costs, the risk of loss of documents and the total time required for the process. It provides a security backup copy of the exam, and can be seen by the pertinent professor and corresponding student (Figure 7).

Figure 7:   Representative view of a scanned image of an exam

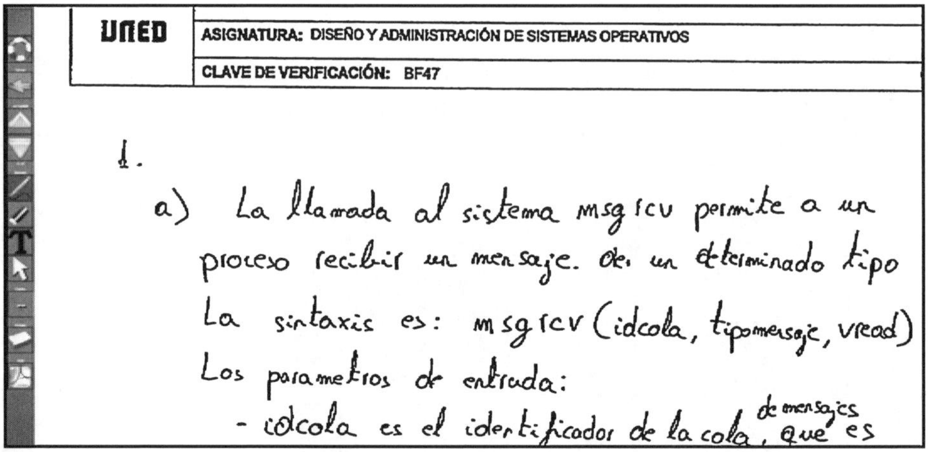

The project started in the 2012-13 academic year. Five centers were selected for implementation at the February call. These were extended to sixteen in June and to nineteen in September. Table 2 shows the number of performed tests.

Table 2:   Number of tests realized in the pilot Centers of Return Bag

Call	Test
February 2013	10,000
June 2013	39,500
September 2013	45,000

To make the correction, the professor has two options: (a) Using the Digital Corrector application that allows on screen correction (PC, tablet or laptop) and export grades to a spreadsheet file for later reading by the Grades Management Application (Figure 8). (b) Downloading the PDF exam files on their computer for later printing.

*Figure 8:   Representative screen of the professors' online database of the corresponding students' exams.
Tools are available for on screen grading or downloading of PDF file for printing.*

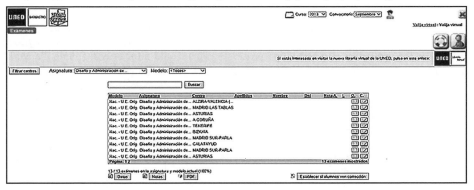

Figure 9 shows the number of exams corrected with the Digital Corrector application. The apparent leveling off in numbers of exams in September is due to the fact that there are only half of the evaluation sessions. Thus, normalizing per week, the number of exams processed with the Digital Correction application is increasingly rapidly.

In 2005, for example, with 110,000 students, about 2.5 million hardcopies of tests were created for distribution by road and air transport to all Associated Centers. That resulted in a cost of 45.000 Euros each call. Nowadays, with 264,059 students, would have implied approximately 3.8 million copies and costs would have exceeded 65,000 Euros per call. Such a cost has been saved thanks to the virtual attaché, which is based in the digital transportation of the exams.

Figure 9:   Evolution of exams digitally corrected during 2013

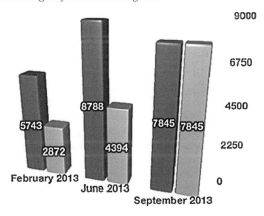

Such an approach has reduces documents transportation costs dramatically. As an example, since 2012, the return virtual attaché has reduced the costs for air transport in 5,000 Euros per call; this is related only to returning the exams from the Spanish islands to the Headquarters in the peninsula.

Further organizational optimizations have been implemented since academic year 2011-12 to reduce the number of professors who have to move from UNED Headquarters to the Associated Center venues. This is based in an increase in the involvement of local collaborators and tutors at local Associated Centers. Savings above 400.000 Euros have achieved due to this. Such a cost optimization has no impact on the quality of service. As conclusions, many economic aspects have been improved with the development of the virtual attaché and virtual attaché return of exams. It is furthermore complementary to the enhancement of security and velocity in the evaluation and correction processes. This would ultimately facilitate the work to the academic and administrative staff and increase the student satisfaction.

After the exam

Shortly after the exam, student may access online to their exam document. The grades of their exam are also uploading in a significantly shorter time. The classical printed procedure demanded almost a month to publish the grades. Due to the characteristics of the virtual attaché, the professors have nearly immediate access to the exams. As a consequence, the time to deliver the grades has dramatically shorted down to an average of one week. Since the student has an access to the grades in such a short time, they may also request access to their corrected exam.This will help them understanding their own performance and draw lessons from the evaluation of the exam. Some students require a certificate of attendance to the exams. The new virtual attaché enables an automatic generation of such a certificate to those students who did go to the exam. This saves time and additional paperwork and time, since this procedure is made online.

Conclusions

A new original system has been developed to simplify the process of preparation, transport, distribution and collection of exams. The so-called "valija virtual" or *virtual attach*é, minimizes the cost and time demand for exam procedures, it eliminates the physical transport of the exams across Spain and worldwide; it eliminates the transportation costs of returning completed exams, which become available to professor and students by secured internet access. The original exams remain guarded at the examination venues. The pertinent professor and corresponding student will have access to the exam. This allows readily access of the professor to the exam for grading it and uploading the spreadsheet with all grades. This approach saves paper and eliminates the conventional method of photocopies; facilitates the work of the evaluation panel and the admission and distribution of students during exam. The new method further enhances the security and velocity in the evaluation and correction processes. Complementary to the virtual attaché return, a new digital corrector program of exams has also been developed for the multiple-choice exams, which has been in use during the last two years. The system is robust. It has been checked and refined during two years with many thousand of exams. It has an important economic cost reduction.

The system is versatile and it has been applied to all types of exams at our university: Access to the University test, Bachelor degree or Master degrees. It is compatible with particularities, such as those in special art documents, pictures or oral exams of different languages. It has also been implemented for specially adapted exams for disabled students.

As a final conclusion, this new method presents a major change in the way to correct, from paper to screen, demonstrating the success of the projects, with very clear advantages. The on-site virtualization endows the university with a more secure, transparent and cost effective evaluation procedure.

Ackowledgements

The authors wish to express our gratitude to all the persons who have been working, during more than 40 years in the evaluation procedures department of UNED. Special mention is devoted to Maria Parrilla-Fernández and Charo Domingo-Martínez pioneers in this Department. We also thanks to E. Jabardo, chief of the Department, M. Albert, S. Alcón and J.A. Merino for their constant interest and attention of every detail on this important process. Our gratitude to M.A. Rodriguez and O. Subiela, virtual attaché coordinators, and to R. Marcos-Urgel and A.I. Rodriguez-Tena. Finally, we thanks to the Fundación Ramón J. Sender and Dr. Carlos Gómez-Mur, Head of Barbastro Associate Center of UNED for the facilities and fruitful development of the new system for virtual attaché.

References

Anderson, T. (2003). Modes of Interaction in Distance Education; in M.G. Moore & W.G. Anderson (2003) (Edits). *Handbook of Distance Education*. New Jersey: Lawrence Erlbaum Associates, Inc., Publishers.

Barberà, E.; Badia, A.; Mominó, J.M. (coords.) (2001). La incógnita de la educación a distancia. Barcelona: ICE-UB /Horsori.

Bernath, U. (2009). Distance and e-learning in transition: learning innovation, technology and social challenges. New Jersey: John Wiley.

Caeiro, S., Leal Filho, W., Jabbour, C.J.C., Azeiteiro, U.M., (Eds) (2013). "Sustainability Assessment Tools in Higher Education Institutions – Mapping Trends and Good Practices Around the World", SPRINGER Springer International Publishing Switzerland 2013. 432 pp. ISBN 978-3-319-02374-8 ISBN 978-3-319-02375-5 (eBook) DOI 10.1007/978-3-319-02375-5 Springer Cham Heidelberg New York Dordrecht London Library of Congress. Control Number: 2013953591 *http://dx.doi.org/10.1007/978-3-319-02375-5*.

Casarotti, M., Filliponi, L., Pieti, L. & Sartori, R. (2002). Educational interaction in distance learning: Analysis of one-way video and two-way audio system. *PsychNology Journal*, 1(1), 28-38.

Cleveland-Innes, M. (2010). An introduction to distance education: understanding teaching and learning in a new era, New York: Routledge.

García Aretio, L. (2011). La educación a distancia. De la teoría a la práctica. Barcelona: Ariel Educación. (4ª ream.).

Holmberg, J., Lundqvist, U., Svanström, M. & Arehag, M. (2012). The university and transformation towards sustainability: The strategy used at Chalmers University of Technology. *International Journal of Sustainability in Higher Education*, 13 (3), 219-231.

Hope, A. y Guiton, P. (2006). Strategies for sustainable open and distance learning. London: Routledge.

Levy, L.M. & Marans, W. (2012). Towards a campus culture of environmental sustainability: Recommendations for a large university. *International Journal of Sustainability in Higher Education*, 13 (4), 365-377.

Litwin, E. (2000). La educación a distancia: temas para el debate en una nueva agenda educativa. Buenos Aires: Amorrortu.

Moore, M., Kearsley, G. (2012). Distance Education: A Systems View of Online Learning (3rd ed.). Belmont, CA: Wadsworth.

Moreno, M. (2012). Veinte visiones de la educación a distancia. México: Edgvirtua.

Peters, O. (2002). Distance education in transition; New trends and challenges, San Francisco: Jossey Bass.

Schlosser, L., & Simonson, M. (2006). Distance education: Definition and glossary of terms (2nd ed.). Charlotte, NC: Information Age Publishing.

Shi, H. & Lai, E. (2013). An alternative university sustainability rating framework with a structured criteria tree. *Journal of Cleaner Production*, 61 59-69.

Simonson, M., Smaldino, S., Albright, M. and Zvacek, S. (2003).Teaching and Learning at a Distance. Foundations of Distance Education. Pearson.

Talbot, Ch. (2004). Estudiar a distancia. Barcelona: Gedisa.

Leveraging E-learning to Prepare Future Educators to Teach Sustainability Topics

Leanna Archambault and Annie Warren[1]

Abstract

As part of the ideals of the New American University, Arizona State University (ASU) has identified the need to work toward efforts to support a more sustainable way of living. Increasingly, we are faced with finding solutions to ensure that the world's population, especially those living in poverty, have the ability to meet their most basic needs. Sustainability definitions articulate this as meeting the needs of the present without jeopardizing the needs of those in the future (Our Common Future, 1986). Education represents the major mechanism for enacting lasting and impactful change.

This chapter describes Sustainability Science for Teachers, a hybrid course in development at Arizona State University that integrates the use of technology and digital storytelling to teach sustainability topics in a meaningful way. This course is required as part of a programmatic education reform aimed at improving science content knowledge among preservice teachers. The goal of the course is for future educators to gain necessary knowledge and skills about sustainability, allowing them to become more informed citizens and helping them learn how to address sustainability concepts within their future classrooms. Not only do future teachers need to understand sustainability topics, but they also need to develop methods for teaching students about these areas and have a grasp of how sustainability fits within the existing curriculum standards (Nolet, 2013). The advantages and disadvantages of employing a hybrid format to teach sustainability topics to future educators are addressed.

Introduction

In today's era, our world is facing significant obstacles in ensuring that all citizens have access to clean water, nutritious food, and secure shelter. Sustainability definitions articulate this notion as meeting the needs of the present without jeopardizing the needs of those in the future (Our Common Future, 1986). Sustainability has become a major concern, and thus a necessary focus, for the 21st century. Considering the increase in the number of people across the globe without access to life's basic necessities, issues of sustainability are more central than ever before. To confront this problem, education is an essential component that must be addressed (Nolet, 2009, 2013). Arizona State University, one of the United States' largest and youngest research institutions, is leading multiple efforts to infuse sustainability as a core element in both its practices as an institution and within its research and teaching agenda. This, in large part, has come as a result of the vision of Dr. Michael Crow, president of Arizona State University. As he sees it, "America's colleges and universities are responsible for the majority of the scientific discovery and technological invention that has advanced sustainability science …. For academic institutions, fostering teaching and research that advances sustainability thus

---

1   Arizona State University, Tempe, AZ USA.

requires new institutional arrangements. But, more broadly, universities should be at the vanguard of producing societal transformation and solutions to the challenges that confront humanity" (2003, p. i, iii).

As a result of this vision, ASU has undertaken several ventures to address this challenge not only in terms of researching and advancing the field of sustainability science, but also through developing programs and curriculum to educate future scientists, scholars, and entrepreneurs. For example, the creation of the Global Institute of Sustainability (GIOS) has brought together a vast array of experts from interdisciplinary fields to share knowledge and work to develop solutions to significant world problems. In addition, ASU has a commitment to use-inspired research and has committed to sustainable practices. This includes installing the largest solar power infrastructure of any institution in the United States, building Leadership in Energy and Environmental Design (LEED) certified structures, and establishing goals of achieving carbon neutrality over the next twenty years. These actions demonstrate a tremendous effort to address current sustainability issues, but are just a step to achieving a sustainable future. One of the most important areas for enacting change is through education of the next generation. For this reason, ASU has created a newly developed hybrid course, Sustainability Science for Teachers, that integrates the use of technology to teach related concepts in a meaningful way. This course is part of the program-wide education reform geared toward preservice elementary teachers that integrates knowledge and skills related to the challenges of improving human health and well-being while reducing human exploitation of natural resources. This course is not only unique in terms of its content and approach to teaching, but also in its interdisciplinary, collaborative approach to the development process. A dynamic team of educators, sustainability scientists, professors, undergraduate, and graduate students, have worked together over the span of a year to create the content, activities, in-class discussions, online videos, and assessments for the course. Elements of the course continue to be developed and refined through an iterative cycle consisting of peer-review and feedback from a variety of sources. This chapter will discuss the Sustainability Science for Teachers course, including its development, its unique hybrid format, and initial lessons learned from creating a course in sustainability geared toward preservice teachers.

Related Literature

Coursework related to teaching sustainability as part of a teacher preparation program is urgently needed (Carney, 2011). Teachers directly shape and influence the next generation of scientists, politicians, inventors, and key leaders – those who will be faced with solving sustainability challenges as these problems become increasingly dire. As part of producing globally-minded and knowledgeable citizens, teachers share responsibility in addressing sustainability literacy. Sustainability literacy is defined as, "the skills, attitudes, competencies, dispositions and values that are necessary for surviving and thriving in the declining conditions of the world in ways which slow down that decline as far as possible" (Stibbe and Luna, 2009, p. 10-11). Themes within sustainability literacy include an intergenerational perspective, the notion of stewardship, social justice and fair distribution, respect for limits, systems thinking and interdependence, the importance of local place, economic restructuring, nature as a model and teacher, and global citizenship (Nolet, 2009, 2013). In addition to these areas, additional issues focus on helping students to appreciate a global perspective and understand that issues, people, and places are interconnected, comprehending how systems function, and being able to think critically to make informed decisions (Church and Skelton, 2009). However, these topics are often not addressed in existing coursework, despite students indicating their eagerness to incorporate them into their teaching (Carney, 2011).

The United Nations Decade of Education for Sustainable Development (United Nations Education, Scientific, and Cultural Organization, 2004) outlined key areas that need to be addressed as part of sustainability education. From this report, a framework for incorporating sustainability education was developed offering recommendations for the preparation of teachers. These guidelines suggest that learning pertaining to sustainability should: (a) embrace an interdisciplinary approach, (b) be embedded within a whole curriculum, (c) involve locally relevant issues, (d) consist of value-driven principles that are explicit so that they can be examined, tested, debated, and implemented; (e) engender critical thinking and problem solving, (f) implement multiple methods and modalities of learning, and (g) view learners as decision makers. These topics have gained prominence as a needed focus among future teachers.

Hybrid Education

From a historical perspective, a number of research studies show that there is no meaningful difference in terms of success, performance, and student outcomes between online and face-to-face learning environments (Cooper 2001, Schulman and Sims, 1999). In a 2000 address, Graham B. Spanier, president of Pennsylvania State University, called, "the convergence of online and traditional face-to-face education the single-greatest unrecognized trend in higher education today". While this may be the case, many scholars are calling for more empirical evidence, case studies, and data that speak to the pedagogical advantages and methodological successes in a clear and succinct manner. The current study provides an examination of one such hybrid course, Sustainability Science for Teachers.

To meet the needs of a growing population, and to be able to reach a wide audience concerning the relevance of sustainability issues, ASU has leveraged the power of online and hybrid education as a way to provide access to digital content to its many students located across four campuses. Higher education has undergone significant change with respect to technological developments that have increased the rapid exchange of information. These advancements in information and communication technologies have led to an evolving landscape across universities, resulting in curriculum delivery methods that are challenging traditional instructional paradigms (Parry, 2010). Educational futurists such as Nathan Harden and Clayton Christensen predicted all facets of education would be changed as we moved into the 21st century (O'Donoghue et al., 2001), and this has increasingly becoming the case. The rise of the part-time, non-residential, non-traditional, lifelong learner is changing higher education from a traditional "brick and mortar" knowledge system to a "clicks and mortar" system (Colucci and Koppel, 2010; Perry, 2010). This transition is fostering new possibilities for content delivery methods from mixed-use multimedia in the classroom, to fully developed learning management systems for online degree programs. While extensive research has been conducted examining fully online education, less is known about the affordances and challenges of the hybrid format.

The hybrid format is a unique blend of two worlds: the face-to-face content delivery method with the all-online course format. The result is a hybrid course, "whereby students participate in learning related activities that extend beyond the boundaries of teacher direction and formal classes" (Martin et al., 2013, p. 51). Bruner (2006) notes that the, "hybrid course has been quietly gaining acceptance as researchers note its structural advantages and pedagogical effectiveness" (p. 229). A hybrid format consists of online content, virtual communications, and interaction from both the student and the instructor. Likewise, it also utilizes in person interactions, the classroom social setting, and typical learning pedagogies to supplement the online learning environment (Heinze and Procter, 2006; Senn, 2008). While there is no formal division or setup for this type of instruction, most utilize a 25-50% reduction in face-to-face meeting times at universities (Dziuban et al., 2005).

A hybrid course format attempts to incorporate the social and face-to-face interactions of the classroom with the convenience and affordances of online learning by implementing specific scheduled face-to-face meetings during the duration of the course (Bruner, 2006; Castle and McGuire, 2010; Lewis, 2010). Currently, few studies focus specifically on the hybrid format for delivering course content in higher education settings (Bruner, 2006; Castle and McGuire, 2010; Lewis, 2010; Senn, 2008). A detailed understanding of the pedagogical connections between engagement with the content and student satisfaction with the hybrid format is desperately needed (Castle and McGuire, 2010; Colucci and Koppel, 2010; Bruner, 2006). Advocates and scholars using the hybrid format denote major advantages that provide a particular cadence that support pedagogical effectiveness (Bruner, 2006). According to the a recent meta-analysis comparing online, hybrid, and face-to-face education, the hybrid format has growing interest in the field as a way to leverage the affordances of both online and face-to-face settings and combining both approaches represented a larger advantage over face-to-face instruction (U.S. Department of Education, 2010). In addition, hybrid formats offer instructors more opportunities to meet students' needs effectively by utilizing a vast array of teaching methodologies and pedagogies (Orhan, 2008).

Sustainability Science for Teachers

Sustainability Science for Teachers is part of a program-wide education reform geared toward preparing preservice elementary teachers to teach students about the challenges of improving human health and well-being while reducing human exploitation of natural resources. Embracing these principles, along with reform movements that move toward additional content-area courses for preservice teachers, a new hybrid course, Sustainability Science for Teachers, was developed and fully deployed in the fall of 2012. The goal of this course is to develop sustainability literacy among preservice teachers by providing the expertise and technology needed to confront seemingly insurmountable societal issues, and to enable future educators to take the concepts they learn and implement them in their classrooms.

Table 1:   Major themes for SCN 400: Sustainability Science for Teachers

Big Idea	Description
We are seeking solutions to problems	Sustainability is about achieving a society where people's needs are met now and in the future. No human need is being met globally in a sustainable way now, so each need becomes a problem seeking a solution.
There are limits to resources	Meeting human needs requires resources that come from nature. There are limits to all resources and therefore there will be limits to the number of people whose needs can be met.
Advances in technology can reset the limits	Developments in technology represent one possible approach to sustainability challenges.
We are dealing with complex systems	All natural systems have many components that interact and one of them, humans, is quite unpredictable.

Big Idea	Description
Solutions require collaboration	We must borrow insights from many fields of expertise to understand nature and our interaction with it. Successful solutions can only be achieved if all the stakeholders are involved.
Problems and solutions are both global and local	Natural cycles, human managed systems, and transportation assure that actions taken at a local level will ultimately have global consequences.
Solutions must be fair	Compliance to policies that achieve sustainable solutions depends on treating everyone fairly.

Three main areas of literature were considered, broaden, and deployed when developing this course for preservice teachers: (1) Environmental Education (EE) which was developed in the mid 20th century and made popular in the 1970's and still used in US schools today, (2) Ecological Literacy as developed by Orr (1989) which calls for an understanding of natural systems and human interactions, and (3) the evolving ideas of Sustainability Literacy constructed from the two previous areas of study in the last two decades. The course developers also took into consideration causality and complexity of science, technology, and society (STS) related to sustainability literacy and how topics of STS are embedded in human relations with and decisions about the Earth's natural and manmade systems (Solomon and Aikenhead, 1994). To consider this connection, emerging epistemologies associated with complexity (Bateson, 1972; Maturana and Varela, 1980) and definitions associated with science and sustainability were also utilized (Kates et al., 2001). Using the domains of environmental education, ecological literacy, and sustainability literacy, major themes were identified as extremely relevant for the course materials. These include the following overarching "big ideas" (Table 1).

These seven major ideas appear weekly in the online content, assignments, and discussions. These big ideas were reconceptualized and streamlined from existing sustainability literacy (Stibbe and Luna, 2009), sustainability competency (Wiek, 2011) and sustainability development literature (Sachs, 1997, 2004) for the preservice teacher audience. These seven themes were decided upon and reviewed carefully by the development team prior to any course materials being fleshed out and fabricated. These themes are not overtly present to the students each week, but appear on a consistent basis throughout each week of the course. Instead, they offer the development team and course instructors a sort of check and balance to ensure materials are consistent in their messages.

Using these seven major themes, the content of Sustainability Science for Teachers was outlined into 14 weekly topics including sustainability, population, poverty, food, water, fossil fuels, new energy, ecosystem services, biome stories, production, disposal, governance, translation, and change. The weekly topic themes were conceived in a specific order that addresses human welfare and the ability to meet basic needs such as food, water, shelter, and economic security first. The approach of tackling global problems and local contextualized solutions of these immediate problems offers the necessary foundation to explore others topics, like biodiversity loss and global production systems which pose potential future threats to humanity. This order is commonly expressed in textbooks that address sustainability, such as *Living in the Environment: Principles, Connections, and Solutions* (Miller, 2013).

Table 2 describes each of the weeks, including the overarching question, and a brief description.

Table 2:   Description of Weekly Topics for Sustainability Science for Teachers

Week	Topic	Overarching Question	Brief Description
1	Sustainability	What is your definition of sustainability?	Students are introduced to the definition of sustainability, the overview of the course, and common myths associated with sustainability.
2	Population	How many people can the Earth support?	Students follow the trials of the human journey from the slow population growth of our ancient ancestors to the population explosion of the last few decades. Together they discover factors affecting why/how populations grow along with the impacts of this growth.
3	Poverty	What does it take to meet everyone's basic needs?	Students hear stories from Armenia, Kenya, Cambodia, and New Delhi. They discuss basic needs, resource distribution, and the extent to which the Millennium Development Goals are being achieved in low, middle, and high-income countries.
4	Food	How sustainable is our food system?	Students discover how food is produced and the consequences for human and environmental health. They explore the problems and solutions associated with our food supply and the management of this system.
5	Water	How can we provide water to meet human needs sustainably?	Students learn about water and sustainability through the lens of three case study countries: Kenya, China, and the US. Students consider the problems and solutions associated with the complexity of water from sanitation to use.
6	Fossil Fuels	How do fossil fuels affect people?	Students learn about our dependence on fossil fuels as a source of energy. They examine the impacts of oil, natural gas, and coal from a broad range of case studies including North Dakota, Louisiana, and Pennsylvania, as well as South Africa and Nigeria.
7	New Energy	What is the future of energy production?	Students explore several types of renewable energy sources. Wind, solar, tidal, biomass, and geothermal power are explored through the social and technological lens. Students are asked to consider the problems and benefits associated with the complexity of each energy type.
8	Ecosystem Services	How strategic is our management of the biosphere?	Students track the integration of human and environmental systems into one "coupled human-environment system." They examine some of the successes of that process, as well as some of the negative effects and consider our relationship with the natural world.
9	Biome Stories	How are humans affecting other life forms on earth?	Students learn about biomes, the benefits of having species diversity, and what efforts are underway to preserve and conserve that diversity.
10	Production	How do systems of production and use affect people and places?	Students learn how the production of goods occurs on both local and global scales, purchasing decisions create positive and negative impacts, and the way we use goods is as important as what we purchase.
11	Disposal	How is waste managed, and how does it affect people and places?	Students learn about the full product life cycle as a system. They also discover that there is a decision to make at the product's end-of-life and that decision can create both positive and negative impacts.

Week	Topic	Overarching Question	Brief Description
12	Governance	How do we create effective policy to govern our actions?	Students learn how science can influence policy by studying the AIDS epidemic, the tragedy of the commons, and a case study regarding the successful reduction of CFCs to decrease the depletion of the ozone.
13	Translation	How will you create a sustainable future?	Students discuss the barriers and solutions to infusing sustainability in their future classrooms.
14	Change	Why does sustainability matter for teachers?	This week students seek to answer the question, "Why does sustainability matter for teachers?" They will consider how teachers can be change agents in their schools and communities, how sustainability can be integrated across content areas using current academic standards, and why sustainability education matters for the future.

Online Content

The weekly online videos use an engaging, documentary-style narrative approach to explore the multiple facets of each topic, taking into consideration the global and national issues of sustainability in 10-minute segments that span a total of 60 minutes per topic. These video narratives employ the use of digital storytelling, a captivating strategy that uses spoken text combined with digital content such as images, illustrations, video, and multimedia to create an emotional connection to the topic (Educause, 2007). Robin (2008) describes the seven essential elements of digital storytelling created by the Center for Digital Storytelling (Table 3).

Table 3:   Seven Elements of Digital Storytelling.

Element	Description
Point of View	What is the main point of the story and what is the perspective of the author?
A Dramatic Question	A key question that keeps the viewer's attention and will be answered by the end of the story.
Emotional Content	Serious issues that come alive in a personal and powerful way and connects the audience to the story.
The Gift of Your Voice	A way to personalize the story to help the audience understand the context.
The Power of the Soundtrack	Music or other sounds that support and embellish the story.
Economy	Using just enough content to tell the story without overloading the viewer.
Pacing	The rhythm of the story and how slowly or quickly it progresses.

The online videos developed for the Sustainability Science for Teachers course incorporate the elements of digital storytelling to present differing viewpoints and perspectives, which is particularly important when dealing with complex sustainability issues. Leveraging the power of story and narrative, the online content is able to convey difficult topics in a more understandable and relatable manner. According to Spierling (2002), "Stories have been used to transfer not only historic knowledge among the generations, but also to transfer cultural and social values to provoke emotions" (p. 2). It is this connection to beliefs and values that is an important part of understanding sustainability challenges and possible solutions. For this reason, the use of digital storytelling is the driving mechanism through which the online content is delivered throughout the course, enabling preservice teachers to connect with sustainability content on a personal level.

In addition to the use of digital storytelling to teach various topics, preservice teachers complete various tasks associated with this content such as quiz questions and written reflections. They also participate in innovative assignments as part of the learning activities. These take a variety of forms such as starting a letter writing campaign, working with non-profit organizations, voting through purchase decisions (conscious consumerism), community action, exploring volunteer opportunities, and developing simple sustainability tools for use in the classroom. The goal of the activities is to encourage students to implement the content they are learning both as informed citizens and as future educators.

Face-to-Face Content

The weekly discussions take place after students have explored the online content. Once per week, small face-to-face discussion sections inspired by course work and weekly topics are led by individual instructors. During the face-to-face sessions of the course, preservice teachers participate in hands-on activities related to the week's topic and explore ways that the content can be implemented in their future classrooms. The in-person sessions offer students the ability to ask questions about the online content, discuss the week's topic with their instructor, and work together to apply how sustainability issues fit within their existing curriculum.

When the content was developed by the team, it was quickly realized a hybrid format would be able to maximize subject area expertize and ensure consistent content delivery by a range of instructors. This was a unique need given the interdisciplinary nature of the course. For these reasons, Sustainability Science for Teachers became one of two required hybrid courses in the preservice teacher education program at ASU. The hybrid format offered the ability to reach students across ASU's four campuses as well as to provide a reliable and balanced presentation of the content. Beyond the delivery of the content, the hybrid format offers students the ability to track their own learning with a variety of assessments such as quizzes and reflections, as well as opportunities for review and reiteration of course materials. The development of this course is very much a work in progress, and revisions continue to be made on an ongoing basis. The format has provided the opportunity to refine the curriculum as part of an iterative cycle in order to improve the course based on student and instructor feedback.

Method

Sustainability Science for Teachers is explored as a case study to identify the advantages and disadvantages of the hybrid format from the perspective of the students and instructors.

Participants

The participants in this study consist of 15 preservice teachers within Mary Lou Fulton Teachers College who were required to take SCN 400. Because these students travel through upper division course work as a cohort, they typically have all classes together until they graduate. Therefore, a very social atmosphere arises from attending and participating in a cohort of learners. Utilizing this social environment in our favor, as part of the course, students in one section had a small group meeting to discuss the course with the researcher. The goal of holding these small group interviews and having students complete short questionnaires was to capture reactions to the course materials for research and development purposes. Students were asked to describe their perceptions of the positive and the negative traits associated with both aspects of the hybrid format, specifically the online component and the face-to-face instruction.

In addition, the instructors that teach the course are also participants in this study. These instructors range from full time tenure-track faculty to part-time faculty associates (FAs). Instructors are identified to teach this course based on the following qualities: understanding of the course content, classroom management techniques, and willingness to thoroughly engage with students to ensure success and mastery of the content. Five doctoral students in two degree programs, four from the School of Sustainability, one from the Human and Social Dimensions of Science and Technology, and three instructors from the Teachers' College comprised the group of instructors for the current study. All instructors were encouraged to participate in the survey to ensure future ideas, concerns, and current materials were reflected upon adequately and captured for both research and development purposes. Instructors were asked about their own perceptions of the advantages and disadvantages associated with the hybrid learning environment as far as teaching was concerned. For example, "From an instructor's perspective, what are the positives and/or advantages of the hybrid learning environment for you as an instructor? What are the negatives and/or disadvantages of the hybrid learning environment for you as an instructor?"

Data Analysis

Using a grounded theory approach, the researchers distributed open-ended questionnaires and conducted interviews of students and instructors designed to elicit perceptions related to advantages and disadvantages of the hybrid format. The resulting data was then coded. Codes were constructed after considering the data and implementing an iterative process. Then the data were reviewed to see what themes were consistent. These consistent and interacting themes were constructed and are the result of the analysis process (Strauss and Corbin, 1998). Interview data were categorized according to dominant themes and coded for analysis. It should be noted, however, that the results are from a limited number of participants who either taught or took the course. Because of the small number of participants and the nature of qualitative inquiry, these findings are not generalizable to other contexts. Rather, the purpose of this study was to explore the perceived advantages and disadvantages of the hybrid format.

Results

Both surveys and interviews yielded similar results. Of the 15 students surveyed, 60% identified the hybrid format as the best learning format for the students in this sustainability course. Likewise, all of the instructors who were interviewed and who responded to the questionnaire noted the advantages of hybrid format. Tables 4 through 7 summarize the findings from the students and the instructors regarding the advantages and disadvantages of the hybrid format used to teach Sustainability Science for Teachers.

Table 4:   Advantages of the format as described by the students' questionnaire responses

Percentage of Response, n=15	Code	Example- Correlated Student Response
30%	Supports learning	"It supports differentiated learning."
40%	Engaging and informative	"In class sessions are more productive because background knowledge has been developed."
100%	Flexible, own pace	"We can learn at our own pace."
30%	Predictable and organized	"You know what to expect each week."
30%	Uses technology in a meaningful way	"It incorporates online technology!" "Videos are so informative and awesome."

Table 5:   Disadvantages of the format as described by the students' questionnaire responses

Percentage of Response, n=15	Code	Example- Correlated Student Response
20%	Cheating	"People can cheat."
20%	Misinterpretation of online materials	"Instructions can be misinterpreted."
30%	Repetitive and/or tedious format	"Repetitive – I don't feel like I am being challenged in having to do the same activities weekly."
30%	Technology issues or conflict with technology personal preferences	"Potential for technology to go bad." "I wish more than one module was open at once."

Table 6:   Advantages of the format as described by the instructors' questionnaire responses.

Percentage of Response, n=8	Code	Example- Correlated Instructor Response
75%	Instructor as facilitator	"Teachers can then focus on facilitating the analysis and application of that knowledge."
40%	Hybrid creates a new student-instructor dynamic	"The instructor gets to build a relationship, in-person, and conduct hands on activities in groups that are more similar to K-8 learning environments."
100%	Engaging format and pedagogy	"Hybrid format capitalizes on the existing knowledge of other experts and conveys complex information on sustainability in an engaging way."

Table 7:   Disadvantages of the format as described by the instructors' questionnaire responses.

Percentage of Response, n=8	Code	Example- Correlated Instructor Response
30%	Lack of instructor-student face time	"Negative is that I just don't get to see my students as often and build a community of learners."
40%	Lack of content control	"One disadvantage is the inflexibility and discreteness of assignments."
80%	Technology issues/Too much self-directed work/Cheating	"Technology issues and waiting till the last minute to complete the hybrid portion. You cannot rush through this and expect to do really well."

Three key findings from the interviews support the surveys. First, both students and instructors thought that the hybrid format was the best of both worlds, incorporating online together with face-to-face elements. Both also believed that hybrid instruction offered a diverse platform for a variety of learning, including the ability to individualize instruction. Finally, both instructors and students believed that a hybrid format created a much more engaging space to learn in. In interviews, students noted the ease of "gaming the system" and cheating on quizzes. During the interviews they elaborated in detail about the issues with the technology and ideas they had to make it better. Students noted their prior experience with online and face-to-face classes and having seen all formats offered at the university, believed the hybrid presented a distinct blend of the positive aspects associated with the different learning environments.

Interviews with the instructors were similar in nature and elaborated beyond what was explored in the written surveys. Instructors indicated that they liked the consistent format that the online portion of the course was able to convey via digital storytelling. It also allowed instructors to focus more on their role as a facilitator able to focus on applying the complex sustainability content knowledge to the K-8 classroom.

As one instructor put it:

The most positive aspect is an organized curriculum that is accessible to all students and appears to be engaging for all students. Many of the topics cannot be fully understood without EXPERIENCING what is happening – which would be hard in an all face-to-face setting. Another positive is the online platform itself as a running record for keeping track of student responses and commenting on them. Laying out the weekly videos and assignments is also very useful to an instructor -it alleviates so many of the (when are things due? What are we suppose to do?) questions.

Another instructor conveyed:

SCN 400 covers a lot of material. Relying on instructors to know that breadth of material and teach it effectively would not be feasible. The hybrid format capitalizes on the existing knowledge of other experts and conveys complex information on sustainability in an engaging way. Teachers can then focus on facilitating the analysis and application of that knowledge.

The results from the questionnaires and interviews supported each other and could be considered a way to triangulate the data in order to identify trustworthiness of the results. The questionnaire data proved to be direct, but lacked the rich details the interviews provided. Administering the questionnaires first and subsequently conducting interviews was successful because it provided necessary background details and illuminated areas for further investigation.

Discussion

When the development process for the course Sustainability Science for Teachers began in 2010, the technological frame was a work in progress as we had difficulty deciding how best to deploy the content to meet the need of our preservice teachers. Three clear choices were identified: completely face-to-face, all online, or hybrid. The team involved in the development of the course content and the format did not establish a specific vision until a year into the project. At this point, clear goals were identified and a trajectory was set. The hybrid format was selected and the development team committed to this idea and format. Because this was a new direction for the dissemination of content, the team concentrated on creating a course that was student-centered and focused on clear student outcomes.

As this chapter illuminates, the hybrid format is a distinctive blend of two worlds, the face-to-face content delivery method and the online course format. The hybrid format appears to have advantages and disadvantages, but ultimately combines the positive and negative aspects of each type of learning environment. This was illustrated in the survey and interview results from the students and teachers. Their reactions to the hybrid format were again commentary on the pedagogical advantages and disadvantages of the sum of the parts of the hybrid course, the online, and the face-to-face formats. However, the hybrid format does offer something for every developer, learner, and instructor, truly creating a unique sociotechnical space. Students indicated that they appreciated being able to go at their own pace through complex content, replaying portions of videos as needed. Instructors had more of an opportunity to be a facilitator when teaching the course. As one instructor described, "This [hybrid format] is the best of both worlds. It gives students a chance to explore the content of sustainability on their own and to build background knowledge that is then applied in the face-to-face meetings. The face to face is the opportunity to hold students accountable for the hybrid portion and to connect the content to real world classroom application."

From the instructor perspective, the hybrid format was able to capitalize on the existing knowledge of experts and to convey intricate and complicated information on sustainability in an engaging way. The use of digital storytelling to captivate preservice teachers was one of the main reasons for developing the course in a hybrid learning environment. Given the multiple perspectives on sustainability issues, being able to present consistent content online from different viewpoints through a relatable, story-driven format, was a distinct advantage. From the student viewpoint, the online content was beneficial because it resulted in class sessions that were more productive due to the additional background knowledge that had been developed prior to attending the face-to-face sessions. Students also appreciated the flexibility of engaging with the content at their convenience and the ability to revisit difficult or hard to understand topics as needed.

However, when discussing aspects of pedagogy and demonstrating various strategies for incorporating the sustainability in the K-8 classroom, the face-to-face setting offered clear benefits. These included allowing the instructor to build their relationship with students in person while being able to conduct hands-on activities that model K-8 lessons. Through the combination of both online and face-to-face learning environments, the hybrid nature of Sustainability Science for Teachers was able to leverage the learning affordances of both formats.

With unprecedented access to information globally, advancements in course offerings at universities will continually evolve. The hybrid format certainly appears to offer many structural advantages such as, being able to present a cohesive curriculum on a complex subjects such as sustainability. The rise of the part-time, non-traditional, lifelong learner is changing higher education and empirical evidence is still needed for specific details on how to best educate students of the future. Hybrid education may be well poised to address these growing concerns.

Conclusion

Sustainability Science for Teachers is aimed at both informing preservice teachers as citizens and assisting them with integrating the related topics into their future teaching. Through the use of innovative and emerging technologies, this course seeks to provide the next generation of educators with knowledge and skills needed to address issues surrounding sustainability. With the benefit of having an assembled team to create such a course, as well as the opportunity to refine and perfect it as part of an iterative cycle, Sustainability Science for Teachers a work in progress. Based on the feedback offered by students and instructors of the course, particularly concerning the advantages and disadvantages of the hybrid format, the team has continued to make improvements to both the online content and using the time in the face-to-face sessions effectively. What remains clear is that, as Nolet (2009) so rightly points out, "… [Sustainability education] is a real phenomenon that merits the attention of the education community, particularly those engaged in the preparation of teachers" (p. 436). This course is an important effort toward educating preservice teachers about the significance of sustainability issues and how they can incorporate these topics in their future classrooms. Using the hybrid format has proven to be an effective method in accomplishing this goal.

References

7 things you should know about digital storytelling. (2007). *Educause*. Retrieved from *https://net.educause.edu/ir/library/pdf/ELI7021.pdf*.

Bateson, G. (1972). Steps to an ecology of mind. San Francisco, CA: Chandler.

Bruner, D. (2006). The potential of the hybrid course vis-à-vis online and traditional courses. *Teaching Theology and Religion, 9*(4), 229-235.

Carney, J. (2011). Growing our own: A case study of teacher candidates learning to teach for sustainability in an elementary school with a garden. *Journal for Sustainability Education*. Retrieved from *http://www.journalofsustainabilityeducation.org/ojs/index.php? journal=jse&page=article &op=view&path[]=46*.

Castle, S., and McGuire, C. (2010). An analysis of student self-assessment of online, blended, and face-to-face learning environments: Implications for sustainable education delivery. *International Education Studies: Canadian Center of Science and Education, 3*(3), 36-40.

Church, W., & Skelton, L. (2010). Sustainability education in K-12 classrooms. *Journal of Sustainability Education*. Retrieved from *http://www.journalofsustainabilityeducation. org/ojs/index.php?journal=jse&page=article&op=view&path%5B%5D=21&path%5B %5D=pdf_9*.

Colucci, W., & Koppel, N. (2010). Impact of the placement and quality of face-to-face meetings in A hybrid distance learning course. *American Journal of Business Education, 3*(2), 119-130.

Cooper, L.W. (2001). A comparison of online and traditional computer application classes. *Technological Horizons in Education Journal, 28*(8), 52-58.

Crow, M. (2013). Presidential Perspectives Forward. In M. Fennell & Miller, S. (Eds.). *Elevating Sustainability Through Academic Leadership* (pp. i-iii). Philadelphia: Aramark.

Dziuban, C.D., Moskal, P. D., & Hartman, J. (2005). Higher education, blended learning, and the generations: Knowledge is power: No more. In J. Bourne & J.C. Moore (Eds.), *Elements of Quality Online Education: Engaging Communities*. Needham, MA: Sloan Center for Online Education.

Heinze, A., & Procter, C. (2006). Online communication and information technology education. *Journal of Information Technology Education* 5, 235-249. Retrieved from *http://jite.org/docu ments/Vol5/v5p235-249Heinze156.pdf*.

Kates, R., et al. (2001). Sustainability Science. *Science* 292 (5517), 641-642.

Lewis, G. S. (2010). I would have had more success if …: Student reflections on their performance in online and blended courses. *American Journal of Business Education*, *3*(11), 13-21.

Martin, R., McGill, T. & Sudweeks, F. (2013). Learning Anywhere, Anytime: Student Motivators for M-learning. *Journal of Information Technology Education, 12,* 51-67.

Maturana, H., & Varela, F. (1980). Autopoiesis and cognition: The realization of the living. Dordrecht: Riedel.

Nolet, V. (2009). Preparing sustainability-literate teachers. *Teachers College Record*, 111(2), 409-422.

Nolet, V. (2013). Teacher education and ESD in the United States: The vision, challenges, and implementation. In R. McKeown & Nolet, V. (Eds.), *Schooling for Sustainable Development in Canada and the United States* (pp. 53-67). New York: Springer.

O'Donoghue, J., Singh, G., Caswell, S., and Molyneux, S. (2001). Pedagogy vs. Technocentrism in virtual universities. *Journal of Computing in Higher Education, 13*(1), 25-46.

Orhan, F. (2008) Redesigning a Course for Blended Learning Environment. *Turkish Online Journal of Distance Education, 9*, 54-66.

Orr, D. (1989). Ecological Literacy. *Conservation Biology 3*(4), 334-335.

Our Common Future – Report of the World Commission on Environment and Development (Brundtland Report) (1986). United Nations. Retrieved from *http://conspect.nl/pdf/ Our_Com mon_Future-Brundtland_Report_1987.pdf.*

Parry, M. (2010). Tomorrow's College. *The Chronicle of Higher Education*. Retrieved from *http://chronicle.com/article/Tomorrows-College/125120/.*

Robin, B. (2008). Digital storytelling: A powerful technology tool for the 21st century classroom. *Theory into Practice, 47,* 220-228.

Sachs, W. (1997). What kind of sustainability? *Resurgence,* 180, 20-22.

Sachs, W. (2004). Environment and human rights. *Development*, *47*(1), 42-49.

Schulman, A.H. & Sims, R.L. (1999). Learning in the online format versus an in-class format: An experimental study. *Technological Horizons in Education Journal, 26*(11), 54-56.

Senn, G. J. (2008). Comparison of face-to-face and hybrid delivery of a course that requires technology skills development. *Journal of Information Technology Education, 7*, 267-283.

Solomon, J., & Aikenhead, G. (Eds). (1994). STS education: International perspectives on reform. Ways of knowing in science series. New York: Teachers College Columbia University.

Spierling, U. (2002). Digital storytelling. *Computers & Graphics, 2*, 1-2.

Stibbe, A. & Luna, H. (2009). Introduction. In A. Stibbe & H. Luna (Eds.), *The handbook of sustainability literacy skills for a changing world* (pp. 9-16). Cornwall, UK: Green Books Ltd.

Strauss, A. & Corbin. J. (1998). Basics of qualitative research: Grounded theory procedures and techniques (2nd ed.). Thousand Oaks, CA: Sage.

U.S. Department of Education Office of Planning, Evaluation, and Policy Development Policy and Program Studies Service. (2010). *Evaluation of evidence-based practices in online learning: A meta-analysis and review of online learning studies.* (Contract number ED-04-CO-0040 Task 0006) .Washington, DC: U.S.

United Nations Education, Scientific, and Cultural Organization. (2004). *United Nations decade of education for sustainable development: Draft international implementation scheme.* Retrieved from *http://portal.unesco.org/education/en/ev.php-URL_ID=36025& URL_DO=DO_TOPIC& URL_SECTION=201.html.*

Wiek, Arnim, Lauren Withycombe, Charles L. Redman (2011). Key competencies in sustain- ability: a reference framework for academic program development. *Sustainability Science, 6*(2), 203-218.

The use of information and communication technologies by secondary school teachers for developing a more sustainable pedagogy in Latvia

Dzintra Iliško and Svetlana Ignatjeva[1]

Abstract

This paper presents an analysis of the use of information and communication technologies in a secondary school setting, as well as the advantages of using ICT in education for sustainability. The authors discuss a pedagogy that is equitable with the use of the ICT technologies. A special emphasis is given to the problems and barriers which prevent secondary school teachers from integrating ICT in their teaching, as well as the main reasons why teachers are willing to integrate those technologies in their teaching. Finally, the authors outline some suggestions for teacher trainers on how to integrate ICT in developing more a sustainable teaching, and on how teachers can improve their ICT literacy to improve their teaching skills.

The paper poses some questions on whether the current pedagogical models are compatible with ICT, and if the integration of ICT can indeed improve learning. Can schools integrate the existing pedagogy with new demands of a digital age? Teachers who are trying to improve their teaching are facing a new challenge of how to adopt new approaches and how to integrate new technologies in their teaching. Which factors serve a contributing role to a successful integration of ICT in teaching? How do these factors make their contribution? To what extent teachers are comfortable with integration of ICT in their subject matter. These questions are addressed in this chapter.

Introduction

The gap between pedagogy and technology is increasing. Schools are being pressed with the new agenda to build 21st century skills into the curriculum, even though not all teachers have adopted sustainable ways of conducting teaching in an increasingly digital world. Schools need to re-evaluate the existing unsustainable pedagogical approaches by integrating ICT technologies into traditional pedagogy (Hedberg, 2011; Labbo, 2006; Zhu and Baylen, 2005). The majority of schools whose pedagogy is based on transmissive and non-constructivist methods of teaching, are trying to integrate new technologies without disturbing the use traditional methods, thus avoiding a renewal of pedagogy. Schools are often being equipped with the necessary materials, but inadequate teacher training during graduation and in-service training, as well as re-direction of teaching methodologies at the school level, causes various problems. The challenge is to ensure that pedagogy can really benefit from new technologies. There is a need to re-engage the field of educational technology with the educational philosophy in mind, and as fitting with learning goals and the subject itself, as well as benefiting teachers and pupils alike (Hedberg, 2011; Hennessy et al., 2005; Koster et al, 2012).

---

1   Daugavpils University, Latvia.

This requires some resistance to traditional approaches within the discourse of 'educational modernization' or 'knowledge society', in which the goals of education remain unquestioned. Schools need to adapt a new pedagogy to the advantage of pupils, which bear in mind the new reality of a networking society.

Towards a more sustainable pedagogy

The way current teachers are teaching, is influenced by various factors such as their experience of teaching and learning. Contemporary schools need to respond to the needs of the present generation, though traditional pedagogical approaches do not longer respond to the needs of a new generation (Christensen and Raynor, 2008; Vrasidas and Glass, 2005). This requires a radical shift in formal pedagogy, towards new epistemologies (Das, 2012). Learning does not occur only in the classroom, and the school is not the only place to access knowledge any more.

Young people live in a constantly connected virtual environment. The array of devices they use is diverse, including smartphones that have been ignored by teachers, in spite of their educational potential. Teachers have deficient training in the field if ICT and this make it harder for them to implement new technologies. Therefore, schools are being left behind by a rapidly changing society, with all its regressive consequences. Teachers need to be aware of the educational benefits and risks of using ICT in teaching. Therefore, it is not enough for teachers to transmit knowledge: they need to become familiar with new ways to teach in a digital world.

ICT provides numerous opportunities for the modernization of the learning process. The integration of such technologies in teaching requires changes in pedagogical approaches, as well a transition from teacher-directed towards more constructivist modes of teaching. Unfortunately, not many teachers new integrate technologies within the framework of traditional teacher-centred paradigms in education in Latvia. Consequently, the use of ICT's becomes merely the aim on its own end. Teachers are working on an environment where promotion and salaris are ofen linked to results; therefore the use of ICT is oriented towards greater achievements instead of developing students' autonomy to work in problem-oriented ways, to cooperate, and to develop self-directed modes of learning.

Kangro and Kangro (2004) argue that, in the humanities, ICTs were integrated as "*add-in*" – an additional model to the existing structure and practice of teaching languages in Latvia. The pedagogical aim of use of e-mails and internet was oriented towards improving the use a foreign language and the awareness of a foreign language in a multicultural setting (p. 34). The main 'value added' purpose of the use of ICTs in Latvia is to facilitate learners' and teachers' professional activities, as well as accessing data, communication, planning and presentation of school work.

Kangro and Kangro (2004) point to an increasingly positive tendency of integrating ICTs for teaching languages. Financial support obtained from EU projects has become another contributing factor in developing ICT infrastructures at schools for teaching science in well-equipped classrooms. Still, ICT is not used as a tool for deeper purposes of learning.

The "*Technology adaptation*" (Niederhauser and Perkmen, 2010) model does not explain all the constraints and motivation of use of technologies in the classrooms. It covers only the access and the availability of an ICT infrastructure dimension in schools, and describes ICT merely as a tool for enhancing teaching rather than explaining deeper levels of motivation of teachers for integrating technologies in teaching (p. 436). Teachers often use technology by building it into existing methods, instead of optimally utilizing technologies with an educational purpose in mind. Technology fails to create a meaningful experience for both, leaners and teachers.

The '*integration model*' of ICT in teaching is more sustainable, since it offers the opportunity to use constructivist approaches, collaborative team work, networking and self-directed modes of learning. It supports the '*technology acceptance*' model developed by Davis (1989) that is supported by a number of external and internal factors. The 'Integration model' allows to view all factors that foster and restrict the use of ICT in schools.

Among the external enabling factors to be mentioned, are the supporting culture at schools, and the support of school administration (Allan et al., 2003), as well as the availability of ICT infrastructure. The culture of schools involves dominant values, such as a shared vision by all teachers, administrator's acceptance of integration of ICTs, and the schools' vision regarding ICT's. Among the internal factors are psychological factors involving teachers' interest (Niederhauser and Perkmen, 2010), attitudes (Tezci, 2009), literacy, knowledge, willingness to integrate technologies in their classrooms, and self-confidence (Davis (1989). Increasing research has been devoted to the use of ICT by teachers, but research on teachers' experience in validating technologies in their teaching has been largely unexplored. As Labbo (2006) argues, while integrating technologies in teaching, educators need to be prepared to negotiate multiple realities that shape their practice.

The use of ICT allows a viewing of the curricula as unfinished, and forces one to re-evaluate the roles that teachers play and the ways they teach. Another issue to consider is how new pedagogy interacts with the institutional policy and teachers' literacy to use ICT.

Methodology of the study

In this study, the authors used both qualitative and quantitative methodologies for the data collection and processing. Quantitative data was gathered by means of a questionnaire applied to the secondary school teachers from four regions in Latvia. The questionnaire includes demographic data (teacher's age, experience of work, field of work) and consists of twenty seven indicators on the purpose and the scope of use of ICTs in teaching designed as Likert-type statements, to which respondents responded with rating from never, rarely, sometimes, often, and always. To determine the content validity of the statements, we piloted them with four experts. The quantitative responses were analyzed by using descriptive statistics. The obtained data were coded and analyzed by the use of the statistical analyses software SPSS Statistics 19.

The statements that derived from the literature and focus group interviews were arranged as bi-polar semantic differential statements, with the purpose to offer teachers a chance to evaluate whether they prefer using ICT or whether their believe in their own authority to develop students' meta-cognitive thinking in supporting students' experience and individual styles of learning without the use of ICT. To ensure the validity of data, the authors used triangulation, by arranging four focus group interviews with a sample of secondary school teachers. Since teachers' beliefs are one of the most important factors when considering a change in their views regarding the role of technologies in teaching (Ertmer, 2005), the authors used focus group interviews to explore teachers' experiences. Qualitative data was obtained in four focus group interviews (N = 63) that was also piloted for reliability with a group of teachers. The pilot group of teachers consisted of participants from the diverse subject areas matriculated in a Master's level programme in education.

The sample

The sample of the study consists of 415 (N=415) high school teachers. Among them, 189 teachers hold a Bachelors degree and 175 are holders of a Master degree. With regard to a branch of their work, 163 of them are teachers of science and 181 are teachers of humanities. This sample of teachers consists of 234 teachers who work in the secondary school setting. 72 teachers are from the secondary schools, and 45 of them are from vocational schools. The study included teachers who work both in city schools and in suburbs. The Mean of teachers' age is 40, and the mean of experience of work is 13, 5 years.

Results of the study

The results obtained from the analyses of data on teachers' attitudinal components toward the use of ICT represent the following findings.

The analyses of indicators allowed a distinguishing of three main groups of factors constraining the use of ICT within the teaching process: *the availability of ICT infrastructure in schools, the scope of use of available infrastructure in the learning process, as well as the efficiency and the purpose of use of ICT in teaching* (Table 1).

The Kaiser-Meyer-Olkin Measure of Sampling Adequacy-KMO obtained is equal to 0,858, which signifies the purposefulness of the use of factor analyses of the designed questionnaire. Factor analyses present three factor structure of teachers' use of technologies in the secondary classroom settings. A factor analyses allowed the authors to single out the following factors:

- Availability of ICT infrastructure in school (F1);
- The scope of use of ICT in the educational process by the teacher (F2);
- The efficiency of use of ICT by the teacher in the educational process (F3).

The most frequently factors can be described by the following indicators:

- The use of ICT infrastructure is promoted at school (0.744, F1);
- Projector is available at school (0.744, F1);
- I use ICT in order to diversify the learning process (0.723, F2);
- I encourage students to use specialized internet forums to promote their ability to formulate their questions correctly (0.817, F3)

Table 1: Factors of the use of ICT by the secondary school teachers

Factors	Indicators	Indicators of factor loadings		
		F1	F2	F3
Availability of ICT infrastructure in school(F1)	The use of ICT infrastructure is promoted at school	0,744		
	Projector is available at school	0,744		
	School has ICT infrastructure	0,695		
	The necessary software is available at school	0,678		
	Interactive whiteboard is available at school	0,655		
	Computers' classroom is available at school	0,624		
	Visual aids are available at school	0,577		
	Computer is available at school	0,562		

Factors	Indicators	Indicators of factor loadings		
		F1	F2	F3
The scope of use of ICT in the educational process by the teacher(F2)	I use ICT in order to diversify the learning process		0,723	
	I use ICT because I believe it helps students to acquire the subject better		0,686	
	I use word processing to create electronic learning materials		0,653	
	I use presentation application to create electronic visual materials		0,653	
	I use ICT in order to show students the opportunities it offers		0,652	
	I look for the necessary information in the Internet search engines		0,540	
	I use the Internet to find tasks for my subject		0,540	
	I use spreadsheet application to perform different calculations and data analysis		0,473	
	I use projector in my lessons.		0,465	
Efficiency of use of ICT by the teacher in the educational process (F3)	I encourage students to use specialized internet forums to promote their ability to formulate their questions correctly			0,817
	I encourage students to use ICT to develop their research skills			0,688
	I encourage students to contact me electronically, thereby motivating students to ask questions more often and when appropriate			0,677
	I use ICT in my lessons in order to promote student ICT skills			0,676
	I share my experience to students about how video materials promote their skills in my subject			0,553
	Cronbach's Alpha	0,837	0,819	0,806
	Rotation Sums of Squared Loadings (% of Variance)	16,8%	16,3%	14,6%

The data presented in Table 1 indicates the percentage of dispersion that describes each of the factors. The highest dispersion (16, 8%) can be explained by the first factor F1. Factor F2 signifies 16, 3% out of total dispersion, F3 is – 14,6%.

Reliability means the repeatability of the measure as well as a consistency of questions. In order to measure the reliability of a questionnaire and its internal consistency, the authors have used the Cronbach's Alpha measurement. Cronbach's Alpha ranges from 0,806 to 0,837 and signifies quite a high internal consistency among the questions included in the questionnaire.

The indicators of factors can be obtained as a result of sum of indicators as well as a result of exploratory factor analyses. The first method is simple from the point of view of implementation, but it does not measure the differences in the loading of several indicators. In the other case, all indicators are being measured, including those loading is insignificant (Field, 2009).

The mean value of all factors is measured by the first method that does not signify large differences from the second method. Besides, the first method of analyses can be used without the use of any special programmes.

For the better understanding and deeper interpretation of data, there was a need of standardization of the value of factors. The standard value of dispersion ranges from 0-1. It is used to measure diverse indicators or the same indicators among the diverse groups of respondents. The standardized meaning of factors allows one to classify respondents in each factor, depending on how high or low are the measures regarding the medium value.

The data collected indicate that the teachers who come from the schools that are equipped with the latest ICT infrastructure, do not use the whole range of possibilities offered by the ICT technologies. The scope of the use of technologies is quite narrow, including only a white board, internet, and use power point. City schools are located in a more advantaged position, and are equipped with better and a more advanced infrastructure than country schools. The whole culture of city schools that is oriented towards innovation and change, influences teachers' decision to make changes in their classrooms as well.

Table 2:   Groups of schools according to a cluster analyses

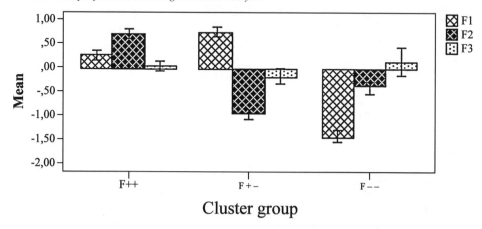

The cluster analysis (Tables 2 and 3) allows to single out three groups of schools:

− F + + schools. The schools that are equipped with the latest technologies; teachers use a wide range of technologies available, but do not make a good use of all the array of technologies available in schools;
− F + − schools. These schools are well-equipped with all the latest technologies but teachers don't use all the possibilities offered by technologies, as well as don't see the efficiency of use of technologies in their classrooms;
− F − − group of schools. The group of schools that are not equipped with the latest technologies, array of available technologies is very poor, and the efficiency of use of ICT is very low either.

The teachers from the group of F++ schools who work in the field of social and nature science comprises 56,4 % of all the teachers. Those schools are well equipped with the latest technologies due to the engagement of school principals, and by the availability of funding from EU projects and governmental programmes. These are mainly city schools (70%) where teachers still do not make good use of all the possibilities offered by the ICT technologies; neither they evaluate the efficiency of use of technologies quite high. This makes one to conclude that the availability of technologies does not necessarily means the purposefulness of its use as viewed by the teachers to a full extent. The teachers from the country schools are more prone to make a good use of existing technologies, rather than teachers from the city schools.

Table 3: Groups of clusters

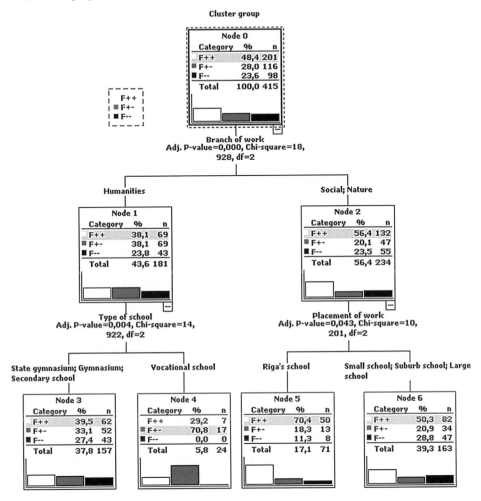

Among the second group of teachers from F+- schools (78%) are teachers who are employed in a vocational school setting that aren't equipped with the latest technologies, display the highest indicator in regards the scope of use of ICT technologies. This can be explained by the specificities of teaching in the vocation schools. In the vocational schools teachers try to make a good use of all the opportunities that technologies can offer.

The teachers from F − - group comprising (23, 5%) of all respondents, are from the field of humanities and (23.8%) and teachers who are employed in the field of Science and social sciences. They do not make a good use of all available technologies at school, as well as they do not see the efficiency of use of ICTs; particularly those teachers, who work in humanities (23%) and in the vocational school setting (0%). City schools (11%) that are the most equipped with the latest technologies do not support teacher's motivation to make a good use of those technologies. This group of teachers does not consider the efficiency of use of technologies in their work either.

Efficiency of use of ICT in learning: data analyses via bi-polar semantic differential

The analyses of data gained via bi-polar semantic differential singled out three main aspects of the purpose of the use of ICT: the role of ICT in promoting students' meta-cognitive level in learning, support of students' experience and individual styles of learning, as well as in promotion of learners' autonomy and socialization (Table 3).

Teachers were asked to evaluate if the use of ICT fosters a more efficient development of students' autonomy, socialization, and meta-cognitive level in learning.

Teachers stressed that they are more efficient in promoting learners' meta-cognitive level of learning, in respecting students' styles of learning and experience without ICT rather than with the help of information technologies. Even though some teachers do not find it comfortable to relieve control and to provide more autonomy on the side of learners, they seem to be willing to consider using information technologies that allow students to proceed at their own pace and to socialize in a digital space. Many teachers feel lost in implementing constructivist modes of learning that provide more control on the side of a learner. As Hennessy et al. (2005) argue, ICTs play a crucial role in fostering pupils' responsibility, autonomy and ownership in doing things, particularly when teachers' competency with ICT is limited, and where pupils are more 'self-sufficient" (p. 174). As autonomy concerns students' feeling of being in control of their own actions and competence in a virtual environment. According to the study, many teachers do not perceive the role of ICT's in developing students' meta-cognitive learning either; they do not see the use of ICT's in respecting students' experience and learning styles either (Figure 1).

Figure 1: The purpose of use of ICT as viewed by the secondary school teachers

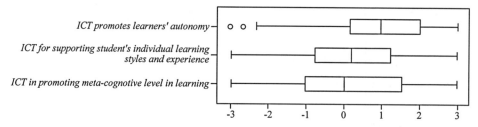

Data gained during focus group interviews

In order to obtain a richer picture on the use of ICT in schools, the authors used four focus group interviews. The authors did not perceive the group of teachers who participated in this study as a homogeneous group, but rather as individuals with different attitudes, beliefs, and basic knowledge about ICT. During the focus group interviews, there was a place for both optimistic and pessimistic rhetoric in regards to e use of new technologies.

Optimistic rhetoric

Among the enabling factors for integrating ICT in schools, the sampled teachers have mentioned a number of external factors, such as **organizational factors**: supportive culture at school, availability of ICT infrastructure, administrative support, proper in- service training; internal factors, such as **psychological factors** that included teachers' interests, attitudes, knowledge, willingness to integrate technologies in their classrooms, and self-confidence.

Teachers' attitudes and beliefs

This study singled out a number of personal factors of computer use. These included teachers' openness to change, recognition of a potential of using new technologies, as well as developing ideas, trying them out, and cindering a specific context. This also entailed how the use of technologies fits into existing subject practices and paradigms. Teachers who claimed to resist integrating of ICT in their teaching, did it out of reasons of practical wisdom. Some teachers are not willing to integrate new technologies in their classes, unless there is a good reason to do so. If teachers do not see a benefit of integrating ICTs in their teaching, they are not willing to change their traditional ways to doing things. Some teachers reported the use of technologies as time consuming and disturbing: *"If technologies don't work it disrupts the whole lesson."* In line with a number of studies (Ruthven, Hennessy, & Deaney, 2005; Wilkan & Molster, 2011) there is a need to emphasize that teachers who adopted constructivist teaching, find the use of ICT useful in motivating students' learning.

ICT for facilitating deeper and more motivated learning

Teachers see a great potential of technologies helping students to see the interconnectedness in subject areas. According to teachers' views, internet and computer software are the most favourable tools for learning languages, developing cooperation and communication skills. One teacher has made an observation: *"Even weaker students become more enthusiastic about learning when they have an assignment to use Internet and mobile phone as a resource"*. One teacher has warned of a danger to overuse technologies: *"If I use a PowerPoint every day, this will be of no use and hamper the learning outcome."* Focus group interview data clearly indicated that teachers recognized the purpose and the congruence of technology use with their own subject.

Supportive culture at school

As teachers admitted, the integration of ICTs at school to a great extend depends on a vision and a motivation of a school's principal to introduce school – wide innovations and inquiry based culture into a school policy. New learning projects facilitated by the variations of instructional strategies, are accepted as typical in innovative schools by the teachers, but perceived as time-consuming in traditional schools. Teachers from the institutions whose whole culture is oriented towards change and innovation, get influenced by this culture and are ready to reshape their pedagogical approaches. One teacher commented that she needs to be sure that ICTs will serve their right purpose of increasing the learning outcome before integrating technologies into teaching. This points to a culture of certainty that still is dominant in many schools in Latvia, and teachers' reluctance to experiment and allow some elements of ambiguity in their teaching. Therefore, as Dawes (2001) argues, there is a need to build the whole school level community of practice associated with the use of ICT's more effectively, regularly and consistently (p. 187).

Pessimistic rhetoric

Among the obstacles mentioned by the sampled teachers are the following: *fear of technologies, old established ways of teaching with a chalk and a board, conservative nature of the traditional culture of schooling, availability and location of technologies, lack of time, limited access, reliability on technology, lack of motivation, lack of technology infrastructure ad an ongoing support, as well as the age factor.* Among the other obstacles for integrating ICT in teaching, many teachers have mentioned lack of training (particularly older teachers), lack of appropriate software in schools, lack of basic knowledge, lack of time in already overloaded schedules, and a lack of technical support at school.

Availability, access, and location of technologies

The situation with regards to technologies in different schools is quite diverse. In some private and city schools, the ICT infrastructure is quite advanced and offers numerous possibilities for teaching, while in country schools the scope and availability of technologies and access to the internet both in schools and at students' homes is very limited. During a focus group interview, many teachers argued about the disparity of availability of technologies in schools and at home. More often in cities, students do have more advanced personal technologies at home compared to those that schools can offer.

They pointed out that the scope of technologies used by the students is quite wide, and are not considered by the teachers as useful teaching resource for the integration in the learning process. The evaluation of all the possibilities and risks that devices involve, the use of ICT's may enhance learning process. In focus group interviews teachers mainly refer to negative impacts of technologies: "*students carry with them all kinds of electronic devices and this is quite disturbing .... I catch students using devices for cheating during tests, forwarding the homework.*"

Age factor

Elderly teachers tend not to use technologies at their lessons. This was reported in the interviews. Quite often, even their attempts to integrate technologies during the classes were not efficient. Their attempt to use technologies is only for the sake of fashion. One teacher made a remark: "*If I use ICT, I try to use it in a pedagogically meaningful way. I need to be clearly aware how and why I use technology, and what aim does it serve*" Unprofessional use of technologies by the teachers diminishes the quality of the lessons. Many teachers have never used ICT based learning strategies as learners themselves, neither had access to proper training or previous experiences of teaching with such technologies that makes the challenge for these teachers even more difficult. As reported by the teachers with bigger teaching experience, computers are being used only for simple purposes, like preparing presentations, e-mails, search of information in internet, while more creative ways of using ICTs are left unexploited.

Technophobia

Elderly teachers trained in the soviet schools system consider that they have an efficient training, and they are very patriotic and responsible about what and how they teach. They would like to use technologies in their classes but they have not developed sufficient skills to use technologies: "*it takes so much time for me to learn things that younger people can learn so quickly*". In line with John's (2005) research, it was discovered that teachers from the humanities are less open for the integration of ICTs in teaching and express more reluctance as compared with the teachers from technology favouring subjects, such as mathematics and science. One teacher has admitted that the problem is with teachers, by saying: "*The less competent is the teacher to use technologies, the more critical becomes the teacher to integrate technologies in teaching*"

Time constraints

As one of the teacher's responded, "*there are many possibilities that technologies can offer. This requires time to learn the latest advancements… The reality is that I am overwhelmed with numerous tasks, paperwork; I need to conduct classes and take care of my family. Yes, IT are engaging for me and my students but I cannot spend so much time on this*"

Conclusions

The integration of technology in teaching is a complex process shaped by teachers' personal pedagogical psychologies, curricular requirements and teachers' attitudes. ICTs should be perceived as an integral part of the shift towards collaborative learner- directed learning. IT is not just 'know-how subject', but ''know-why subject' (Seeman, 2003). Technology education needs to be learner-centered, by dealing with the issues that are relevant to the learner's environment by making the learning experience of the students of central significance and interdisciplinary in approach.

Among the main obstacles mentioned by the teachers that keep them resistant to integrating technologies into teaching, are the following: insufficient time they spend to acquire the latest technologies, phobia to use technologies, lack of ICT infrastructure in schools, and a lack of time (digital technologies are time consuming to learn and to implement, needing greater time-investment than many teachers can afford).

ICT can be used to enhance participatory learning, interdisciplinary teaching, holistic and interactive ICT-based teaching. Technologies can be integrated for the pedagogical enrichment of teaching. Teachers need to be facilitators in "mixing digital activities and non-digital ones, mixing presence and distance, dealing with time and space" and all the possibilities offered by the ICT (Cornu, 2012, 17).

The major question is how to integrate new technologies in the learning process, so that they can be beneficial to students' leaning in providing richer contexts. Technology is a tool, but the teachers play a crucial role in re-developing schools into technology enhanced institutions.

The integration of technologies in schools does not follow a single model; the process is highly variable from school to school and is determined by array of factors. Schools need to respond to a new generation in various ways. The transforming potential of technologies marks a clear shift from the delivery of contents to constructivist modes to learning and knowledge production (Gibbs and Gosper, 2006; McLoughlin and Lee, 2008).

Suggestions for teacher trainers for training teachers to use ICT

Creation of sustainable pedagogical approaches and models that are relevant to the demands of the 21st century, is an imperative of a day. If schools are unable to do so, they will be left behind (Hargreaves, 2003). Teachers need to become life-long learners, by re-examining the skills needed in a digitalized world to be able to handle new technologies that serve a meaningful purpose.

There is no single sufficient answer to the success of integration of ICT in teaching; numerous teachers' narratives reveal a set of conditions for a successful integration of ICT in teaching. Teachers do not integrate ICT in their teaching unless they are fully able to do it, if they are equipped with support and are engaged in a meaning making process in a way that technology informs their practice in a creative way. The integration of ICT in teaching requires teachers to accept uncertainty and complexity of introducing new approaches and strategies.

In-service teacher training programmes need to enhance teachers' ability to explore and to experiment. Teachers need to explore all the possibilities that the use of ICT can offer them. Also, teachers need to be encouraged to experiment with new and adapted pedagogies appropriate for digital technologies. Teacher training institutions need to perceive teachers not as a homogeneous group, but rather be aware of the need to search diverse ways in supporting them to respond to the challenges of the day.

Schools should become a supportive learning places, where teachers can engage in a technology-enhanced learning and to develop more sustainable pedagogies. Schools should avoid isolating technology as a separate discipline. Instead, they need to demonstrate how technology can support sustainable pedagogy.

The integration of ICT in teaching requires teachers to acquire and to develop diverse pedagogical strategies, for facilitating the handling of subjects in a way that caters for more effective classrooms interaction. The new roles to be played by teachers in incorporating ICT in teaching, are complex and demanding. This requires access to knowledge on new strategies for mediating interaction between ICT and pupils. Universities need to identify what kind of support teachers need, so as to meet the new demands current and new technologies pose.

It is not enough to simply list the obstacles to be overcome on how to integrate ICT in teaching. A continuous support to teachers is required, and examples of successful practice of teaching with the technologies are needed. Technology should not be considered as a separate unity, but as integrated part of teaching process, and the teacher himself/herself needs to assess the appropriateness of technology use as matched with the aim, the purpose and the methods of a lesson. Apart from this, teachers need to be confident with the technology they choose to use.

ICT teaching must be part of teachers' professional development and they must be given time to evaluate and to experience if the use of ICT can indeed enhance learning.

References

Allan, H. K., Law, N., and Wong, K.C. (2003), "ICT implementation and school leadership: Case studies of ICT integration in teaching and learning," *Journal of Educational Administration*, 41(2), 158-170.

Christensen, C.M. & Raynor, M. E. (2008), Disrupting class: How disruptive innovation will change the way the world learns. New York: McGraw Hill.

Cornu, B. (2010), "ICT in teacher education: Policy, open educational resources and partnership." In *Proceedings of the UNESCO Institute for Information Technologies in Education*, St Petersburg: Russia, November 15-16, pp. 12-17.

Das, S. (2012), On two metaphors for pedagogy and creativity in the digital era: liquid and solid learning. *Innovations in Education and Teaching International*, 49(2), 182-193.

Dawas, L. (2001), What stops teachers using new technology? In M. Leask (ed), *Issues in Teaching Using ICT*. London: Routledge, 61-79.

Ertmer, P. A. (2005), "Teacher pedagogical beliefs: The final frontier in our quest for technology integration?" *Educational Technology Research and Development*, 53(4), 25-39.

Field, A. P. (2009), *Discovering statistics using SPSS*. London, England : SAGE.

Gibbs, D. and Gosper, M. (2006), "The upside–down–world of e-learning." *Journal of Learning Design* Vol. 1, Iss. 2, pp. 46-54, Retrieved from *http://jld.quit.edu.au./*

Hargreaves, A. (2003), Technology in Knowledge Society. Education in Age of Insecurity. New York: Teachers' College, Columbia University.

Hennessy, S., Ruthven, K., and Brndley, S. (2005), "Teacher perspectives on integrating ICT into subject teaching: commitment, constraints, caution, and change." *Curriculum Studies*, 37(2), 155-192.

Hedberg, J.G. (2011), Towards a disruptive pedagogy: changing classroom practice with technologies and digital content. *Educational Media International*, 48(1), 1-16.

John, P. (2005), "The sacred and the profane: Subject sub-culture, pedagogical practice and teachers' perceptions of the classroom uses of ICT," *Educational Review*, 57(4), 471-490, *DOI:10.1080/00131910500279577.*

Kangro, A. and Kangro, I. (2004), "Integration of ICT in Teacher Education and Different School Subjects in Latvia," *Educational Media International*. Routledge: Tailor & Francis Group, DOI: 10.1080/0952398032000105076, pp. 31-37.

Koster, S., Kuipert, E., and Volmant, M. (2012), "Concept-guided development of ICT use in 'traditional' and 'innovative' primary schools: what types of ICT use do schools develop?" *Journal of Computer Assisted Learning*, 28, 454-464.

Labbo, L.D. (2006), "Literacy pedagogy and computer technologies: Towards solving the puzzle of current and future classroom practices," *Australian Journal of language and literacy*, 29(3), 199-209.

McLoughlin, C. and Lee, M. (2008), "Future learning landscapes. Transforming pedagogy through social software," *Innovate,* 4(5), p. 7.

Niederhauser, D. S. and Perkmen, S. (2010), "Beyond self – efficacy: Measuring pre service teachers' instructional technology outcome expectations." *Computers in Human Behaviour*, 26(3), 436-442.

Ruthven, K., Hennessy, S. and Deaney, R. (2005), "Incorporating internet resources into classroom practice. Pedagogical practices and strategies of secondary-school subject teachers," *Computers and Education,* 44, 1-34.

Seeman, K. (2003), "Basic principles in holistic technology education," *Journal of Technology Education,* 14 (2), 28-39.

Tarle, P. (2004), "A theoretical and instrumental framework for implementing change in ICT in education." *Cambridge Journal of Education,* 34(3), 331-350. *DOI: 10.1080/03057640420 00289956.*

Tezci, E. (2009), "Teachers' effect on ICT use in education: The Turkey sample." *Proceedia Social and Behavioural Sciences,* 1, 1285-1294.

Vrasidas, C., Glass, G.V. (Eds). (2005), Preparing teachers to teach with technology. Greenwch, CT: Information Age Publishing.

Wikan, G. & Molster, T. (2011), Norwegian secondary school teachers and ICT. *European Journal of Teacher Education,* 34(2), 209-218.

Zhu, E. & Baylen, D. M. (2005), From learning community to community learning: pedagogy, technology a interactivity. *Educational Media International,* 42(3), 251-268.

Appendix 1: The use of ICT technologies in secondary schools

1. Gender
   - Male
   - Female

2. Age
   - _____

3. Work experience
   - _____

4. Education
   - Currently studying
   - Bachelor's degree
   - Master's degree
   - Doctor's degree

5. Branch of work
   - Social sciences
   - Natural sciences
   - Humanities

6. Placement of work
   - Suburb school
   - Small school
   - Large school
   - Riga's school

7. Type of school
   - Secondary school
   - Vocational school
   - Gymnasium
   - State gymnasium

	Never	Rarely	Sometimes	Often	Always
I look for the necessary information in the Internet search engines	☐	☐	☐	☐	☐
I use the Internet to find tasks for my subject	☐	☐	☐	☐	☐
I use projector in my lessons.	☐	☐	☐	☐	☐
I use interactive whiteboard in my lessons	☐	☐	☐	☐	☐
I use visual aids in my lessons	☐	☐	☐	☐	☐
I use spreadsheet application to perform different calculations and data analysis	☐	☐	☐	☐	☐
I use word processing to create electronic learning materials	☐	☐	☐	☐	☐
I use presentation application to create electronic visual materials	☐	☐	☐	☐	☐
I share self-made materials on the Internet	☐	☐	☐	☐	☐
Computer is available at school	☐	☐	☐	☐	☐
Computers' classroom is available at school	☐	☐	☐	☐	☐
Projector is available at school	☐	☐	☐	☐	☐
Interactive whiteboard is available at school	☐	☐	☐	☐	☐
Visual aids are available at school	☐	☐	☐	☐	☐
The necessary software is available at school	☐	☐	☐	☐	☐
School has ICT infrastructure	☐	☐	☐	☐	☐
The use of ICT infrastructure is promoted at school	☐	☐	☐	☐	☐
I use ICT in order to diversify the learning process	☐	☐	☐	☐	☐
I use ICT in order to show students the opportunities it offers	☐	☐	☐	☐	☐
I use ICT because I believe it helps students to acquire the subject better	☐	☐	☐	☐	☐
I use ICT in my lessons in order to promote student ICT skills	☐	☐	☐	☐	☐
I encourage students to contact me electronically, thereby motivating students to ask questions more often and when appropriate	☐	☐	☐	☐	☐
I encourage students to use specialized internet forums to promote their ability to formulate their questions correctly	☐	☐	☐	☐	☐
I encourage students to use ICT to develop their research skills	☐	☐	☐	☐	☐
I share my experience to students about how video materials promote their skills in my subject	☐	☐	☐	☐	☐

Appendix 2: The purpose of the use of ICT technologies in the secondary school

	+3	+2	+1	0	-1	-2	-3	
ICT promotes greater interest about the learning process	☐	☐	☐	☐	☐	☐	☐	Teacher promotes greater interest about the learning process
ICT promotes students' belief in their success	☐	☐	☐	☐	☐	☐	☐	Teacher with his/her professional approach towards learning process promotes students' belief in their success
ICT promotes deeper students' cognitive functioning	☐	☐	☐	☐	☐	☐	☐	Teacher promotes deeper students' cognitive functioning
ICT guides students towards self-directed learning	☐	☐	☐	☐	☐	☐	☐	Teacher guides students towards self-directed learning
The use of ICT promotes students' personal experiences	☐	☐	☐	☐	☐	☐	☐	Teacher promotes students' personal experiences
The use of ICT promotes students' social interaction	☐	☐	☐	☐	☐	☐	☐	Teacher promotes students' social interaction
ICT promotes learners' autonomy	☐	☐	☐	☐	☐	☐	☐	Teacher promotes learners' autonomy
The use of ICT promotes diverse learning achievements	☐	☐	☐	☐	☐	☐	☐	Teacher promotes diverse learning achievements
ICT promotes differentiated approach towards learning	☐	☐	☐	☐	☐	☐	☐	Teacher promotes differentiated approach towards learning
The use of ICT promotes students to use diverse solutions for solving their tasks	☐	☐	☐	☐	☐	☐	☐	Teacher promotes students to use diverse solutions for solving their tasks
ICT promotes students to experiment and visualize	☐	☐	☐	☐	☐	☐	☐	Teacher promotes students to experiment and visualize

Appendix 3: Focus group interview questions

Are schools ready to reshape pedagogy and teaching approaches with the demands of a digital age?

Which factors play a contributing role to a successful integration of ICT in your teaching?

How do these factors make their contribution in your teaching?

Which factors prevent integration of ICT in your teaching?

Why do you choose or not to choose to integrate ICT in your teaching?

To what extent are you comfortable with integration of ICT in your subject matter?

Do you believe that integration of ICT in teaching can challenge to become students actively involved in learning? In what way?

Does integration if ICT's motivate your students for learning? How? Please, give examples.

III.
A Continental Perspective for best practices

A critical narrative of e-learning spaces for sustainable development in the Global South

Rudi W. Pretorius[1]

Abstract

Using information and communication technologies (ICT), specifically referring to e-learning, is becoming the accepted norm in higher education and has the potential to support, improve and expand education for sustainability (EfS). The reality, however, is that implementation of e-learning poses several challenges for teaching staff, students and administrators in contexts such as the Global South – a collective name for emerging and developing nations. Although these challenges vary from country to country, limitations in terms of access to the Internet – even among university students – remains a barrier for the effective roll-out of education to exactly those segments of the community needing it the most. Since EfS per definition cannot exclude certain community segments, implementation of e-learning in contexts as the Global South should proceed with due consideration of matters such as the latter. E-learning is indeed not the panacea for all the ills in the educational system. In support of this viewpoint, two narratives of transforming sustainability related modules to e-learning at the University of South Africa (Unisa) are presented and critically reflected on. This chapter provides evidence favouring a gradual, phased approach to the implementation of e-learning. Such an approach suits the context of EfS in the Global South better than would be the case for an abrupt change-over to e-learning.

Introduction

Education for sustainability in the Global South

Although education for sustainability (EfS), viewed as holistic, morally grounded learning focused on concern and skills to address the world's sustainability crisis (Fien, 2001), has been achieving prominence, it is also true that it is still in the background in many regions of the world (Manteaw, 2010). Focusing on Africa, Manteaw (2012) observes that although now past its halfway milestone, the United Nations Decade of Education for Sustainable Development (DESD), and associated with it the idea of EfS, are at large unheard of. Manteaw (2012) maintains that a similar situation exists in many other regions as well, notably in the Global South, defined as countries with a medium to low human development index (World Bank criteria), located in South America, Africa and Asia and comprising about 133 of the 197 countries of the world (Damerow, 2010).

Rigg (2007) explains the importance of the Global South in terms of population (80% plus of world total) and GDP (estimated 60% plus of world total by 2025) and also a range of prevailing political, socio-economic and environmental issues. But the nature of people-environment interactions in the Global South challenges the ability of humankind to explore

---

1   Department of Geography, University of South Africa, Florida, 1710 South Africa.

alternatives for increased sustainability in ways differing from "Northern"/Westernised approaches. To address this challenge, sustainable development needs to be presented meaningfully to different audiences in the Global South, mindful of unique local contexts. This can be done through EfS based on appropriate educational philosophies and practices, linked with community based epistemologies (Manteaw, 2012).

The potential of e-learning, including some challenges

It is acknowledged that the number of people in need of better access to educational opportunities is much higher in countries forming part of the Global South, than for those in the "North" (McNaught, 2005). This holds true for the need for access to programmes in EfS as well, which Manteaw (2012) refers to as "sporadic" and "incoherent" for the African context. Open and distance learning (ODL) potentially has the ability to respond to this need and in a relatively fast way as well. If combined with the use of information and communication technologies (ICT's), this becomes even more viable. As pointed out by McNaugth (2005), there are only a few places left in the world that are by now not touched at all by the impact of the global information web.

As the Internet infrastructure is becoming more stable and more affordable in parts of many countries in the Global South, e-learning is more and more considered and implemented as delivery mode to facilitate educational opportunities. Despite this, the lack of reliable basic infrastructure (such as electrical and communications supply and the relatively high cost of bandwidth), continues to limit not only economic growth but also expansion in the educational sector (Fay and Morrison, 2007; Unwin et al., 2010; Wright et al., 2009). Noteworthy is that technology-led innovations frequently fail to produce the transformations anticipated by lecturing staff (Kirkwood and Price, 2008), which can be linked to (among other things) improper exploitation of the pedagogic potential of ICT's.

Aim of chapter and methodology

The aim of this chapter is to map, contextualize and critically reflect on the transformation from paper-based to online offering of two sustainability related modules at the University of South Africa (Unisa). Unisa is a major ODL provider in the Global South, and currently embarking on a large scale roll-out to create an increased e-presence for its course offerings. The offerings of the Department of Geography are currently in various stages of transformation to e-learning, of which the two modules being reflected on in this chapter, forms part of. In this chapter the adoption of a gradual, phased approach to implementation of e-leaning in the one module (undergraduate) is mapped and then compared with the immediate and direct change-over to e-learning in the case of the other module (postgraduate – honours).

This mapping and comparison is presented in the form of a narrative reflection by the author, who is directly involved with the modules being considered. This narrative format is closely related to the technique of reflective writing, as set out by Jasper (2005). The value of and need for narrative enquiry as qualitative analytic strategy in educational research is highlighted by McNaught (2005). In this chapter a reflexive position is taken to consider the strategies being used and the experiences of the various role players. The two narratives include reference to the personal experience of the author as well as the author's perception of how students are experiencing e-learning. Challenges as well as possible ways to deal with these are presented as well as suggestions on the best possible route to follow for institutions planning to implement e-learning.

Unisa: Institutional context and e-learning

National level

Unisa is the only dedicated ODL higher education provider in South Africa. With a head count of more than 300 000 in 2012 (Unisa, 2012), Unisa is not only the biggest university on the African continent but also in South Africa, and is counted amongst the significant examples of ODL institutions in the world (Tait, 2008). After a merger exercise between government subsidised ODL providers in South Africa, a new institution emerged in 2004, adopting "Unisa" as name again, but now with vision *"Towards the African University in the service of humanity"*. In terms of not only its size, but also its aggregated resources, experience and capacities, Unisa is in a unique position to provide crucial inputs in terms of EfS in Southern Africa and even beyond (Unisa, 2005).

Since 2005 Unisa implemented numerous technologically based strategies, to increase the e-presence of its academic offerings. This initiative has been constrained by issues concerning access to ICT's by many students in South Africa (Oyedemi, 2012). Using 1044 students as sample, Oyedemi (2012) indicates that 62.9% of them do not have Internet access at home. Access is rather from computer labs on campus (used by 60.3% most regularly), then mobile phones (used by 47.1% most regularly) and then from home (used by 29.6% most regularly). Oyedemi (2012) highlights that for the different population groups in South Africa, 76,9% of Black students do not have Internet access from home, compared to 43.2% of Indian/Asian students and 11.4% of respectively Coloured and White students. In terms of provisioning of EfS to students as comprehensively as possible, the referred to variations are problematic and needs to be dealt with with care.

Institutional level

Unisa is positioned as a comprehensive, open distance learning institution with mission to produce scholarship and research, to provide quality tuition and to foster meaningful community engagement (Unisa, 2011a). Sustainability is embedded in the vision of Unisa (Unisa, 2007) and provides an indication of Unisa's sincerity in addressing Africa's environmental, developmental, social and educational challenges (Unisa, 2011b). Unisa therefore forms part of the world wide move among institutions for higher education to implement policies, operations and curricula in support of and towards greater sustainability.

Initially the switch by Unisa to technological based strategies was necessitated by the inadequacy of all systems (administrative, operational and tuition) to cope with ever increasing student numbers, together with externalities such as an ineffective and unreliable national postal system. Examples of the switch to technology include online student applications and registrations, online access to brochures and registration information, the *my*Unisa virtual learning environment and the online submission and onscreen marking of assignments. These initiatives obviously have greening implications, and assist in creating a suitable milieu for EfS.

Departmental level

Over the past 30 years the Department of Geography at Unisa went through a number of re-curriculation exercises. As a result the compartmentalized sub-disciplinary focus was transformed to an approach using issues manifesting on various scales in the world as structuring mechanism (Pretorius, 2005). Together with this a structured study program in environmental management was introduced in 2000, with the teaching style in the Department of Geography transformed to be more student-centered and to foster more active involvement of students with their studies (Pretorius, 2004). Through these initiatives the Department of Geography contributed to not only pioneering EfS in the institution but also to support the worldwide sustainability agenda.

Most recently the Department of Geography embarked on the transformation of its offerings to be offered through e-learning. Noteworthy in this regard is the re-design of the Geography Honours degree, which has been offered exclusively through e-learning the first time in 2013. This degree is part of a small group of degrees that are the first to be offered exclusively online in the institution, an achievement that the Department of Geography is proud of. In addition, the Department of Geography also offers one module in the group of so-called "signature modules" (the first undergraduate modules in the institution going online), namely "Environmental Awareness and Responsibility". From 2013 all new students have to include at least one of these signature modules in their degrees.

Personal level

The author witnessed, participated in and took the lead in many of the changes in the Department of Geography over the past almost 30 years. When arriving in the Department of Geography in 1987 (Pretorius, 2004), the state of art was encapsulated by content driven teaching of the sub-disciplines of Geography through first generation distance education. Good about this was the perfected model of "wrap-around" study guides, the recognition obtained for good first generation distance education materials and the high standards in general. Bad about it was the lack of student-lecturer interaction, the promotion of rote learning, assessment being treated as an after-thought and little room for application and problem-solving.

Over these 30 years, the author's role changed from presenter of modules to that of change agent, not only to influence and guide curriculum changes, but also to facilitate new teaching and learning approaches. The latter involves experimentation with inquiry based, contextual and authentic learning, on which the author presented a few conference papers and also published some journal articles and book chapters. On a different level, the author played a major role in planning and implementing the undergraduate programme in environmental management and is still involved as programme coordinator. The current challenge is to successfully convert and continue with these initiatives in terms of the institutional move towards e-learning, which will be highlighted in the narrative accounts now following.

First narrative: Undergraduate module – Geography of tourism

Essentials of the learning experience

"Geography of tourism" is a 2nd level module focusing on the interaction between people (tourists) and the environment (destination areas), with reference to the physical, economic and socio-cultural impacts of tourism, as well as related sustainability and management implications (Pretorius, 2008). This module has been designed in such a way as to enable students to develop their own voice concerning the geography of tourism and to apply it in appropriate contexts. The outcomes to be achieved centre around the environment as resource for tourism, the needs and behaviour of tourists on terms of different spatial and time scales, the interaction between tourists and the environment and lastly the sustainable management and utilisation of tourism resources

In line with the requirements for EfS, the learning experience for students is contextualised with the aid of authentic stories, scenarios and case studies. In addition, students have the freedom to apply what they have learnt in their local contexts, or else to use any other context of their choice. This approach is extended to the formative and summative assessment and thus students have the opportunity to apply the theory in real world contexts. This type of assessment assists students to engage in knowledge construction and to take ownership of their own learning. All in all the idea was to get as close as possible to so-called "Rich environments for active learning" (REALs), as originally suggested by Grabinger and Dunlap (1995).

More about the students

Students who enroll for this module are mostly doing either the BA or BSc in Environmental Management or the BCOM in Tourism Management, while some are doing either a general BA or BSc with a Geography major or are busy qualifying as teachers by doing a BEd. During 2013 student numbers were about 120 per semester, which can be regarded as a relatively small class size in the Unisa context. The composition of the class typically varies in terms of age, gender and race, with students located all over South Africa and even a few in other countries as well. Although many of these students already have jobs (i.e. in the tourism or environmental management sectors), just as many lack work experience and started with their studies directly after leaving school. Due to the diverse student body, it is not possible to design teaching and learning strategies focused on any specific group. This diversity, however, creates the potential for lively online discussion and sharing of their experiences from various contexts.

Approach to e-learning

"Geography of tourism" has not been designed and is not offered as a fully online module. The approach being followed can be described as "blended learning" (Walker & Keeffe, 2010) in that various forms of paper based study material (study guide, reader, tutorial letters, etc) are supplemented with a variety of technologically based means, such as video conferencing, text messages to students and activities on the *my*Unisa virtual platform, although as yet these are still mostly optional. Assignments can be submitted either electronically (in which case it is marked onscreen and sent back to students electronically) or in hard copy (in which case it is marked by the traditional "hand method" and then posted back to students).

In 2013 a blogging exercise on *my*Unisa has been introduced as part of the first assignment for this module. It also contributed a small percentage to the marks for this assignment. The purpose of this exercise is to orientate students in choosing an appropriate context for their assessments and also to get feedback on their approach towards the first assignment. Following participation in the blogging exercise, it is hoped that students will be more eager to participate in the discussion forums for their other assignments, although optional. Further activities on the virtual platform include posting of generic feedback on assignments by means of the wiki tool (thus allowing posting of comments) and using the additional resources tool to provide various support materials (such as templates for maps and tables to be used in assignments, feedback on assignments, presentations used in video conferences, etc.).

Lecturer's experience

Taking into account evidence pointing towards the access issue presenting difficulties for many South African students to go online (Oyedemi, 2012), the blended learning approach followed in this module is the author's preferred option. It affords students with or without Internet access the opportunity to be able to do the module. In addition, assignments can be compiled with or without the aid of technology, depending on the individual situation of students. Noteworthy is that experience from 2009 to 2013 indicates that using technology does not necessarily imply higher assignment marks. Manually prepared assignments in many instances presented authentic work, whereas plagiarism from the Internet seems to be a growing problem with electronically submitted assignments.

The introduction of a low-key but credit bearing blogging activity as part of an assignment works quite well, with over half of the class participating during the 2013 academic year. Since the weight of this activity is quite low, not doing it did not hugely affect students negatively. But it serves a valuable purpose in that it gets students to the point of using the virtual platform and realising the benefits. It is a pity though, that student participation in subsequent non-credit bearing activities was not as good. Luring students towards the virtual platform by continuously pointing out advantages of using it, is what it is about. Another advantage is an increase in computer literacy levels of students, thus contributing towards their graduateness. Those not online are obviously missing out in this regard.

Perceptions of student's experience

As experienced in the 2013 academic year, the students that do in fact participate in the online blogging activity and grasped what it actually is about, seemed to benefit greatly. They have used this opportunity optimally to orientate themselves towards the requirements for the module and the first assignment. Positive feedback includes comments on the advantages of being able to communicate directly with a lecturer, to get feedback almost immediately after making a blog post and being able to see what contexts class mates are using for their assessments. Many students, however, just post a blog to get it done, not realising that a return visit is required to read and react on the comments on their blog posts. It is clear that the computer literacy of many of these students is not at the level where it should be.

Noteworthy is that the participation rate of students in the 2013 academic year in subsequent optional online activities was much lower. The message coming from this is that if these activities are not credit bearing, students will not easily do them, even if they are obviously beneficial to do. Concerning students that cannot go online, the message is very negative, indicating feelings of being disadvantaged compared to those students with Internet access. As indicated in the previous section, the reality is that this is not necessarily the case, but this does not affect the perception that students may have in this regard. Once students grasp the benefits of going online a few times and how that may assist them, they are generally less negative about the idea of using the *my*Unisa virtual platform.

Concluding remarks on this narrative

Although the blended learning approach utilised in "Geography of tourism" has many short-comings, it affords the opportunity to those not online yet to gradually get acquainted with the online learning environment. In this way they can be exposed to the advantages associated with it, although structural constraints might still prevent them to participate in e-learning at this stage. The implications of the digital divide, however are stark, with those being online and submitting assignments online generally performing better than their counter-parts that are not online yet. This can be the result of something as simplistic as the longer turn-around time between submission and receipt of feedback associated with the manual submission of assignments compared to online submission. Although the blended approach is suitable for the transition from paper based teaching and learning to a fully online system, it can be foreseen that once access issues and other associated problems have been addressed, that the online model will become the preferred choice, with the bulk of student body accepting and utilising it in that way.

Second narrative: Postgraduate module –
The geography of people-resource interaction in the Global South

Essentials of the learning experience

"The geography of people-resource interaction in the Global South" is a compulsory module in the Geography Honors degree offered by Unisa (junior postgraduate or 4th level of study at South African universities). By means of exposure to a variety of authentic contexts in the Global South, this module guides students to critically reflect on the spectrum of resource issues this region is grappling with, while developing their skills to actively engage with these issues with a view to effect real and lasting improvements. The outcomes to be achieved centre around the ability to conduct a resource analysis, consideration of discourses on resource use, framing of an authentic task concerning resource use and lastly to alert role players about various aspects related to resource use.

The problem based approach to learning that is taken, is aligned with EfS thus allowing holistic consideration and analysis of the unstructured, open-ended, real word problem of interaction between resources and people in the Global South. This problem has many dimensions and students may view it from different perspectives, depending on the contexts they find themselves in. Taking these individual unique contexts as point of departure, students are guided to refine the problem of people-resource interaction in terms of relevance in their chosen contexts. In this way they can set up the parameters to analyse this problem and develop solutions appropriate for their context. To end of the module this is extended to facilitate comparisons with situations elsewhere, and even with the "more developed" parts of the world as well.

More about the students

Students who enroll for "The geography of people-resource interaction in the Global South" have not necessarily done their first degrees at Unisa, but come from a variety of universities in South Africa. All of them are also not necessarily enrolled for the Geography Honors, since students at this level may include one module from a related field in their qualification if they want to. Student numbers at the honors level are generally much lower than for undergraduate modules. In 2013 the class for this module consisted of about 25 "active" students. Since these students are specialising at post graduate level, the diversity of the class in terms of background tends to be less problematic than for undergraduate modules. In addition, the smaller class size makes the facilitation of teaching and learning relatively easy. Although students enrolling for this module are aware of the requirement of Internet access, they are not required to provide evidence about it at this stage.

Approach to e-learning

"The geography of people-resource interaction in the Global South" has been conceptualized as a fully online module and been implemented in 2013. There is no printed study material such as study guides and tutorial letters. Study guidance is via interactive, online learning units, presented in five focus areas. Students are linked to material available online, such as video clips, web pages, journal articles and e-books, also including Open Educational Resources (OER). No specific texts are prescribed, but students are rather guided to search in a dedicated way online for materials relevant to their specific people-resources context. Comprehension of the various dimensions of people-resource interaction in the Global South does not come in the form of neatly pre-packaged prescribed texts.

During the 2013 academic year, formative assessment firstly comprised of a number of credit bearing online activities (structured contributions to blogs, discussion forums and wikis) that students were required to engage with and complete as they worked through the online learning units. A total of 10 of these online activities had to be completed, two for each of five focus areas. In addition, more substantial assignments (taking the form of critical essays, presentations, progress reports, etc) also had to be submitted during the year. These are all submitted electronically and marked onscreen. In total all the formative assessments contributed 50% to the final mark obtained. The remaining 50% comprised a comprehensive e-portfolio submitted at the end of the study period.

Lecturer's experience

Since this is a postgraduate module, preparing students to be able to eventually function as researchers, the lecturer felt comfortable with the idea to offer this module in an exclusive online environment. Despite access issues, it is not possible to provide a quality learning experience to these students if they are not prepared to and are unable to use the Internet to support their study efforts. It seemed, however, that for the 2013 academic year, less than 50% of the students were equipped to function effectively as students and budding researchers in the virtual environment. Surprisingly many students simply ignored the online activities (despite the fact that they were credit bearing) and did not work through the online learning units either.

From the lecture's point of view, it is clear that in the 2013 academic year those students who indeed engaged with the learning units, and did the online activities, seem to have grasped what is required and succeeded in achieving the module outcomes. The opposite however, holds true for approximately more than half of the students who did not participate as required. The issue in this regard might have more to do with the general preparedness of these students for post graduate studies, than whether they are able to cope with e-learning or not. Much less interaction and activity were expected of these students during the preceding paper based era, during which their under-preparedness for post graduate studies was not as upfront as is the case now.

Perception of student's experience

From comments supplied during the 2013 academic year in blog postings, discussion forums and wikis, it appears that although students are not in principal against online learning and actually realise the value of it, many of them are skeptical about their own abilities to keep up in terms of what is required. Despite their relatively positive initial response, it soon became clear that a significant number of 2013 students were doing a very small number of the online activities, if at all, and were also submitting assignments without any evidence of engagement with the online learning units and/or the material being linked to. Apart from the under-preparedness already referred to, the access issue (Oyedemi, 2012) might be playing a role here as well. An example in this regard that the lecturer became aware of is that of a student who had to use 3G via a fellow student's laptop, which is not always available.

Concerning the assessment structure in the 2013 academic year, students seem to feel that there were too many things to do, to the extent that they struggled to find time to keep up with everything. In cases where students are doing three, four or the full set of five modules of the Geography Honors qualification, this can indeed be a problem. All the "small" nice to have online activities add up and this aspect can easily become more important than original-ly intended. An aspect to work on is that although students do their own posts in blogs, dis-cussion forums and wikis, they are reluctant to react on the posts of fellow students. This jeopardises the ideal of peer learning and forming a community of practice. On a positive note, students who grasped what is required and were participating fully, had a much richer learning experience than was possible before in the now bygone paper based era.

Concluding remarks on this narrative

The narrative on "The geography of people-resource interaction in the Global South" firstly illustrates the general under-preparedness of students to participate meaningfully in post graduate studies. This is associated with a relatively low participation rate in online activities, while a large number of students complete assignments without proper consultation of the online learning units. At this stage the access issue can still be regarded as an aggravating factor, with the computer literacy of students generally not at a level conducive to online learning. This is despite the fact that this module has Internet access as requirement. In order to run an online module of this nature successfully, it is clear that a major effort will need to be undertaken to upgrade the computer literacy skills of students and to assist them with the initial use of the various tools available on the *my*Unisa virtual platform. The bewilderness of many students can be ascribed to the fact that they mostly come from undergraduate contexts where tuition was either through contact or paper based mode. As undergraduate students are exposed more and more to online learning, it can be expected that their performance in this regard at postgraduate level will drastically improve.

Implications in terms of e-learning for sustainable development

First narrative: Gradual implementation

In terms of the agenda for EfS, the approach to gradually implement e-learning through blended learning, holds great promise. This is especially true for higher education institutions in the Global South where many students still seem to be battling with amongst other things access issues, unreliability of the Internet infrastructure and the relatively high costs associated with Internet usage. By forcing an abrupt change-over to e-learning, the risk of loosing students, thereby killing there passion for EfS, is just too big. By first giving students exposure to e-learning through the blended approach before switching to a fully online system, therefore holds distinct advantages.

The biggest advantage of the blended learning approach lies in the fact that it has the ability to take the local contexts within which students find themselves, into account. The latter is a valuable advantage of blended learning, as for instance reported by Thorn (2003), and also in line with the requirements of EfS. In addition, blended learning has the potential to combine the innovative technological advances offered by e-leaning, with best practices in traditional paper based distance education. The needs of students who require a flexible approach in terms of time and place are catered for as well, which is another positive in terms of EfS.

Second narrative: Abrupt change-over

The problem associated with an abrupt change-over to e-learning is that the students are not really prepared for it and do not expect it either. As a result they experience the change-over as a traumatic exercise, to the extent that they may even go to the extreme to abandon their studies. In terms of the agenda for EfS, this is of course highly undesirable and should be prevented if at all possible. The frustration associated with the mode of learning should not be allowed to kill the passion of those showing interest to the work as change agents within the field of environmental sustainability.

However, since the world is becoming increasingly connected in terms of information networks as the Internet, the need to be able to work, communicate and find information online is not negotiable anymore. This is especially true at post graduate level, where students are expected to work independently. Since a huge volume of information on the issue of sustainability is available on the Internet, it is unthinkable that graduating students would not know how to go about to obtain it. Therefore, if transferring to e-learning abruptly, support systems in this regard need to be put in place for students in the field of EfS, or else run the risk of frustrating and loosing them.

In conclusion

In terms of the low rate of Internet penetration in many countries in the Global South, including South Africa, a relatively large online absence in courses offered through this mode should be expected. For the two narratives on modules transforming to e-learning that have been presented in this chapter, the fact that about 50% of students and sometimes even more have no online presence is therefore not unexpected, but indeed still worrying. Under such circumstances the implication is that higher education institutions might not be able to fulfill their mandate in terms of education for sustainability, and specifically those potential students who really are in need of educational opportunities. For courses using a blended learning model, and not immediately going fully online, this obviously presents less of an issue. The reality is that structural constraints and societal inequalities associated with a low rate of Internet penetration might not only be difficult but actually impossible to address in the short to medium term. A model according to which e-learning is gradually phased in, might therefore be more preferable to an abrupt change-over in this regard.

References

Damerow, H. (2010), Global South, International Politics (GOV 205), Union County College, Cranford. Retrieved from *http://faculty.ucc.edu/egh-damerow/global_south.htm.*

De-George-Walker, L. & Keeffe, M. (2010), "Self-determined blended learning: a case study of blended learning design", *Higher Education Research and Development*, 29(1), 1-13.

Fay, M. & Morrison, M. (2007), Infrastructure in Latin America and the Caribbean: Recent developments and key challenges, Washington DC, World Bank Publications.

Fien, J. (2001), Education for sustainability: reorienting Australian schools for a sustainable future, *Tela Series*, Iss. 8, Fitzroy, Australian Conservation Foundation.

Grabinger, R.S. & Dunlap, J.C. (1995), "Rich environments for active learning: a definition", *ALT-J: Research in Learning Technology,* Vol. 3, Iss. 2, pp. 5-34.

Jasper, M.A. (2005), "Using reflective writing in research", *Journal of Research in Nursing*, 10(3), 247-260.

Manteaw, B.O. (2010), "Global environmental politics and the discourse of education for sustainable development: Why the discourse needs attention", *International Journal of Environment and Sustainable Development,* 9(1/2/3), 74-90.

Manteaw, O.O. (2012), "Education for sustainable development in Africa: The search for pedagogical logic", *International Journal of Educational Development*, 32(3), 376-383.

McNaught. C. (2005), "Understanding the contexts in which we work", *Open Learning*, 20(3), 205-209.

Oyedemi, T.K. (2012), "Digital inequalities and implications for social inequalities: A study of Internet penetration amongst university students in South Africa", *Telematics and Informatics* 29, 302-313.

Pretorius, R.W. (2004), "An environmental management qualification through distance education", *International Journal of Sustainability in Higher Education*, 5(1), 63-80.

Pretorius, R.W. (2005), "A narrative of change: reflecting on two decades of Geography at the University of South Africa", paper presented at the *Sixth Biennial Conference of the Society of South African Geographers,* University of the Western Cape, Bellville, 5-10 September.

Pretorius, R.W. (2008), Geography of tourism, only study guide for GGH206Y, Pretoria, University of South Africa.

Rigg, J. (2007), An everyday geography of the Global South, Abingdon, Routledge.

Tait, A. (2008), "What are open universities for?", *Open Learning: The Journal of Open, Distance and e-Learning*, 23(2), 85-93.

Thorn, K. (2003), Blended learning: how to integrate online and traditional, London, Kogan Page.

Unisa. (2005), 2015 Strategic plan – An agenda for transformation, Retrieved from *http://unesdoc. unesco.org/images/0014/001486/148654e.pdf.*

Unisa. (2007), Unisa Service Charter, Retrieved from *http://www.unisa.ac.za/cmys/staff/ contents/docs/Final_ServiceCharter_120307.pdf.*

Unisa. (2011a), United Nations Global Compact – University of South Africa – Communication on progress. Retrieved from *http://www.unisa.ac.za/cmsys/staff/contents/docs/ UNGC_Report_ 20032012.pdf.*

Unisa. (2011b), University of South Africa: Charter on Transformation, Retrieved from *http://www.unisa.ac.za/cmsys/staff/contents/docs/Unisa%20Transformation%20Charter %2027072011.pdf.*

Unisa. (2012), An institutional profile of Unisa: Unisa Facts & Figures, Retrieved from *http://heda.unisa.ac.za/filearchive/Facts%20&%20Figures/Briefing%20Report%20Unisa% 20Facts %20&%20Figures%2020120215.pdf.*

Unwin, T., Williams, L. M. O., Alwala, J., Mutimucuio, I., Eduardo, F., & Muianga, X. (2010), "Digital Learning Management Systems in Africa: rhetoric and reality", Surrey, University of London, Retrieved from *http://www.gg.rhul.ac.uk/Ict4d/LMS.pdf.*

Wright, C.R., Dhanarajan, G. & Reju, S.A. (2009), "Recurring issues encountered by distance educators in developing and emerging nations", *International Review of Research in Open and Distance Learning,* 10(1), 1-25.

Cotonou 2012 and Beyond

—

An Assessment of E-learning for Sustainability in sub-Sahara Africa

J. Manyitabot Takang and Christine N. Bukania[1]

Abstract

It would seem that e-learning is finally taking its place in contributing to education for sustainable development in Africa. This chapter offers an assessment of e-learning for sustainability in sub-Sahara Africa, by drawing on some case studies such as the African Virtual University. The assessment will investigate to what extent the e-learning initiatives are home-grown, i.e. specifically designed to solve local problems, how e-learning is bridging the lack of human capacity, and access to information on sustainability. Moreover, it will address the question of how inclusive e-learning for sustainability could be in Africa, especially in view of infrastructural limitations and the costs of participating in e-learning courses. The chapter concludes with some recommendations for future developments in e-learning for sustainability in Africa.

Introduction

Sustainability is about human survival and human survival is about the survival of coupled socio-ecological systems. To underscore this link, it suffices to reiterate the fact that human welfare relies on natural resources such as water and forests, as well as the economic, social and political frameworks within which these resources are exploited. The complexity that underlies this link has serious implications for the study and operationalization of *sustainability*[2]. In fact, sustainability, as a paradigm for the analysis and interpretation of human-nature interactions, has been described as a hybrid, cross-disciplinary, multi-disciplinary field that incorporates almost all other known disciplines (Sterling 2012; Carolan 2008; Boix Mansilla *et al.,* 2000). It comes then as no surprise that whereas sustainability is one of the most used words of modern times, agreement on what it actually is and how it should be implemented is hard to reach. The Brundtland definition of sustainable development[3] remains one of the most widely cited ones. It describes sustainable development as that which *"meets the needs of the present without compromising the ability of future generations to meet their own needs."* This

---

1   Environmental Governance Institute (EGI), Buea, Cameroon.

2   Sustainability has been approached from both a human-centred approach (weak sustainability) as well as from a very eco-centric approach (strong sustainability). See for instance Neumayer, E (2003). *Weak versus strong sustainability: exploring the limits of two opposing paradigms*, Elgar, London.

3   The Brundtland report was among the first comprehensive reports that sought to reconcile development with environmental concerns. Throughout their report, *Our Common Future*, the World Commission on Environment and Development (WCED) use "sustainable development" and "sustainability" interchangeably.

conceptualization of sustainability necessitates the complex integration of economic, social and environmental considerations into development planning. The defining feature of the Brundtland report is its emphasis on the human dimensions of sustainability, that is, underlying beliefs and value systems, attitudes, behaviours and knowledge relating to natural resources and the environment. This emphasis was picked up and elaborated in Agenda 21, the global blueprint for implementing sustainable development. While several chapters of Agenda 21 apportion roles among different parts of society[4], an entire chapter is dedicated to the kind of education that will challenge and reshape our knowledge and beliefs[5].

However, over 20 years since the Brundtland report, the path to *our common future* seems more elusive than ever. Several comprehensive reports have not only highlighted the persistence of business as usual models but also shown that loss of biological diversity continues at alarming rates[6]. But it is perhaps the issue of climatic change that speaks for deteriorating environmental conditions more than any other. If more immediate and concerted action is not taken, climate change threatens to reverse all developments that have been made over the past decades towards a sustainable future.[7] The impacts of climate change will be unequal in different parts of the world, depending on capacities to adapt. For sub-Saharan Africa droughts, decrease in rain-fed farming and increase in water stress have been predicted for some parts, while increased incidence of storms and floods will affect other parts of the continent (Takang, 2012a).[8] With the lowest life expectancy, lowest incomes and lowest school enrolments of all major world regions (UNDP, 2005: 222), such setbacks will not improve conditions around the continent. Quite on the contrary, the prevailing conditions are likely to fuel conflicts over scarce resources[9].

It is now evident that the Millennium Development Goals (MDGs) will not be achieved in their entirety[10]. The MDG Report 2013 reports a steady increase in poverty rates in sub-Saharan Africa between 1990 and 2010 (UN, 2013: 7). In view of these challenges, one cannot overstate the case for education for sustainable development and more so the role that e-learning could play in facilitating this.

In recent times, there has been considerable interest in the role that information, communications and telecommunications (ICT) can play in fast-tracking achievement of global sustainable development (SD). For example, it is now widely acknowledged that access to the

---

4 Several chapters of Agenda 21 are dedicated to the roles that different strata of society should play in achieving sustainable development. See Chapter 23 on the role of major groups; Chapter 24 for the role of women; Chapter 25 on the role of Children and Youth; Chapter 26 on the role of indigenous peoples and their communities; Chapter 27 on the role of NGO; Chapter 28 on Local Authorities; Chapter 29 on Workers and Trade Unions; Chapter 30 on Business and Industry; Chapter 31 on Scientific and Technological Community and Chapter 32 on the role of farmers.

5 Chapter 36 of Agenda 21 titled *promoting education, public awareness and training*, calls for (a) reorienting education towards sustainable development, (b) increasing public awareness and (c) promoting training.

6 Millennium Ecosystems Assessments, UNEP Global Environment Outlook and Africa Environment Outlook.

7 See especially Boko, M., I. Niang, A. Nyong, C. Vogel, A. Githeko, M. Medany, B. Osman-Elasha, R. Tabo and P. Yanda, 2007: *Africa. Climate Change 2007: Impacts, Adaptation and Vulnerability.* IPCC. See also Stern, Nicholas. (2007). The Economics of Climate Change – The Stern Review. Cambridge: Cambridge University Press.

8 See UNEP Global Environment Outlook 4, especially at 207.

9 For a discussion on environmental conflicts, their occurrence and impacts on sustainability see Takang (2012b). *International Conflict Resolution.* The Encyclopedia of Sustainability: Vol. 9. Afro-Eurasia: Assessing Sustainability (pp. 168-172).

10 See the Millennium Development Goals Report 2013.

internet can facilitate access to markets, medical services and weather information to aid farmers in remote parts of the world. According to the World Summit on the Information Society (WSIS), the growth driven by ICTs can contribute to poverty eradication and sustainable development (WSIS 2003, Principles 33, 43). At the global scale, initiatives such as the United Nations Global Alliance for ICT and Development (UN GAID) have underscored the importance of ICT for SD. At continental scale, this thematic was directly addressed by the *e-learning Africa 2012* conference that was held in Cotonou, Benin from 23-25 May, and whose theme was *e-learning and Sustainability*.

E-learning Africa is Africa's most important stakeholder forum when it comes to e-learning and ICT supported education. The conference brings together policy makers, decision makers and practitioners from education, business and governments. The Cotonou 2012 conference explored the role that e-learning can play in fostering sustainable development in Africa. It is against this backdrop that this chapter assesses e-learning for sustainability in Africa.

Many have argued that Africa is on the verge of a technological jump into the twenty-first century, largely driven by the mobile revolution being witnessed on the continent (Isaacs, 2013; Fuchs and Horak, 2008). This technological revolution has attracted investments in infrastructure and increased access to educational opportunities. This notwithstanding, decade-long resource shortages continue to plague the formal education system: poor infrastructure, inadequate staffing and teaching and learning materials paint a drab picture for those who are yet to join the technological superhighway[11]. This dichotomy of the haves and have nots that characterizes the development trends of the continent can also be witnessed in the development of e-learning.

As the United Nations decade for education for sustainable development (2005-2014) draws to a close, it is vital to re-examine the role that education for sustainable development (ESD) can play in the transition from the Millennium Development Goals to the post-2015 Sustainable Development Goals, and particularly how e-learning can be used to enhance ESD on the African continent.

This chapter will assess e-learning for sustainability in sub-Sahara Africa by drawing on some case studies such as the African Virtual University. Different e-learning strategies across Africa will be considered, as well as collaborative initiatives between African institutions and their development partners.

The assessment considers the extent to which e-learning initiatives are home-grown, i.e. specifically designed to solve local problems; and how e-learning is bridging the lack of human capacity and access to information on sustainability. The chapter concludes by providing recommendation for changes in the aftermath of the Cotonou 2012 e-learning conference.

Education for Sustainable Development (ESD)

According to Schultz (1961), education enhances individual options available to people. At the same time, an educated populace provides the skills required for industrial production and economic growth. Therefore, the right kind of education is crucial in changing unsustainable economic and social practices and in bringing the world back to a sustainable development

---

11   This has been described as the digital divide: "inequality of access to the Internet" (Castells, 2002: 248); "the gap between those who do and do not have access to computers and the Internet" (Van Dijk, 2006: 178). See also "an inequality in access, distribution, and use of information and communication technologies between two or more populations" (Wilson, 2006: 300).

path (McKeown *et al.,* 2006). Put in another way, without the right kind of education, the knowledge and skills necessary for sustainable production will not be transferred. Moreover, only the right education will move contemporary beliefs, values and attitudes away from consumerism and unsustainable practices. This kind of education, which in essence embodies a different approach, is what we would term as education for sustainable development. Such an approach hinges on the conception of development as one that goes beyond the improvement of quality of life to ensure greater individual freedom, raise standards of health and nutrition, increase access to and equal opportunity, ensure a cleaner environment and facilitate cultural diversity (Dreze and Sen, 1995). It promotes development that is not limited to income and economic growth, but one that empowers people, and enables them to attain self-satisfaction and self-respect a development that does not just distribute scarce resources but also redistributes proceeds accrued from the use of such resources. In Africa, where the gap between rich and poor is great and marginalisation is real, sustainable development is a social contract within and across human generations[12].

Some have asserted that ESD is an extension of environmental education, whose aim is to foster participatory environmental problem solving (Tilbury, 1995; Hesselink *et al.,* 2002) and to emphasise critical thinking and problem solving skills (Huckle and Sterling 1996). According to others, ESD is education that requires a global systems literacy (Wood, 2012: 505), and promotes a proper comprehension of coupled socio-ecological systems while emphasising local realities[13]. Wood goes further to assert that it will take *interdisciplinarity* to "encourage students to explore connections between traditionally isolated disciplines" (ibid.: 505). Munson (1997) argues along similar lines by stating that ESD is neither a subject nor discipline but permeates all subject areas and all levels of education while, Sitarz (1993: 202) asserts that ESD "*involves how each subject relates with environmental, social and economic issues*".

When these conceptions are considered, it becomes clear that ESD differs fundamentally from mainstream education. It entails education that is entrenched in eco-centric understandings. Unlike conventional education, ESD approaches and pedagogies are as different as the contexts in which they emerge. The goal is however the same, to create a responsible citizenry that can contribute to reducing our ecological footprint. Sustainability is not an end-state but a dynamic shaping and reshaping of social, economic and environmental issues. Interpreted in this way, ESD too must cut across all strata of society, time and space to foster lifelong learning skills.

E-learning and ESD

Holmes and Gardner (2006) define e-learning as the use of new multimedia technologies and the internet to improve the quality of learning by facilitating access to resources and services as well as remote exchanges and collaboration. E-learning can considerably increase capacity development for sustainability in Africa by facilitating access to knowledge, skills and education. The key attributes that make e-learning so appealing include its shift of focus from teacher-centered to learner-centered education; reduction in the cost of education; improvements in access to learning opportunities (Engelbrecht, 2003); and that e-learning neither has time nor space constraints. Flexibility of access and convenience (Cantoni *et al.,* 2004) are

---

12   Intra-generational and inter-generational equity.
13   A doctrine that arose after Agenda 21 is "think global, act local".

the features that make e-learning so suitable for ESD. Due to the large volumes of information on the internet, people, especially in Africa, who lack access to education for geographic, physical or social reasons see their chances improved through e-learning. For these reasons, e-learning has been hailed on the argument that ICT could improve the quality of learning while at the same time improving access to education at reduced costs. However, as Weigel (2002:1) warns, if e-learning programmes do not generate workers capable of solving intricate and authentic developmental problems, then they are not worth much. The way in which the potentials of ICTs for ESD in Africa are harnessed will largely depend on the types of models that are developed and applied. It is against this backdrop that the next section takes a closer look at some examples of e-learning programs across the African continent.

E-learning Programs in Africa

Several models have emerged across Africa in the implementation of e-learning for sustainability. This section will take a closer look at a few of such models.

Open education resources (OER)

As a model, Open Educational Resources (OER) includes educational resources that are openly available to everyone at no cost (Butcher 2013: 30). By virtue of their free content, OERs are open textbooks of sorts and are among the most widely used e-learning models in Africa[14]. Nonetheless, whereas the content is free to the user, there are costs involved in accessing it[15]. E-learning through OER typically constitutes a multimedia document[16] designed for students to use by themselves in self-study. These could be full courses, textbooks, streaming videos, exams, software, and other materials or techniques supporting learning (The William and Flora Hewlett Foundation, 2010).

Massive Open Online Courses (MOOCs)

Massive Open Online Courses (MOOCs) for their part have predominantly featured in discussions on the role of e-learning for sustainability in higher education. According to Rodriguez (2013: 68) MOOCs are global events that are neither limited by space nor degree of openness. In this sense MOOCs create a forum for adapting information and in which teachers take on new roles such as moderator and facilitator (Kop et al., 2011).

Pferffermann (2013) explains that the potential of MOOCs has especially been seen in the role that open courses, especially from renowned international universities could play. Unlike OERs, MOOCs are not necessarily "open". Easter and Ewins (2010:12) point out that most MOOC initiatives in Africa involve external, often global partners. This involvement is based on the premise that multi-stakeholder partnerships for education can encourage innovation

---

14   The Convention to Combat Desertification UNCCD for instance has a "market for capacity building", where different modules are offered. Many of the courses are tailored to developing countries including those of Africa. See *http://www.unccd.int/en/programmes/Capacity-building/CBW/Resources/Pages/E-learning.aspx*

15   Learners would typically require an internet connection, which is not always free.

16   These include free license content e.g. Creative Commons (*www.creativecommons.org*).

and development of context-relevant education, leverage financial resources and improve the learning process.

Some Examples

In the section that follows, some examples of e-learning for sustainability initiatives will be used to illustrate trends across the African continent.

The African Virtual University (AVU)

To begin with a classic example, the Virtual University (AVU) was established in 1997 as a World Bank project. In 2003, it became a Pan African intergovernmental organisation[17]. According to AVU's charter, it has the "*mandate of significantly increasing access to quality higher education and training through the innovative use of information communication technologies*"[18]. Since it started in 1997, AVU has trained more than 43,000 students and has established partnerships with over 53 partner institutions in 27 Anglophone, Francophone and Lusophone countries in Africa. The AVU offers courses using three different modes. Online courses are offered fully through the AVU virtual classrooms, which have individual log-in portals through which students can have access to lecturers, learning resources, live sessions and the digital library, and access technical support when needed. In the online courses, students and lecturers can interact audio-visually in real-time, but can also access recorded sessions. In mixed-courses, compulsory face-to-face lecture sessions are delivered in partnership with partner institutions. For example, AVU has partnered with the Royal Melbourne Institute of Technology and Curtin University in Australia and Université Laval in Canada. In the third mode, all teaching and learning is done face-to-face and the courses made available online.

AVU offers courses with a core focus on sustainability such as renewable energy[19], as well as food security[20], among others.

AVU was conceived and launched in Washington. In its initial two-year phase, lectures were delivered to 15 African countries via video-conferencing from Washington. It was not until 2000 that the AVU was moved to Nairobi, although the lectures continued to be streamed from Washington.

The cost of the programme made it heavily reliant on donor funding, and prevented it from being scaled up to reach more students. Now in its final phase of implementation, AVU has developed a communication and marketing strategy to increase publicity for its course offering. Nonetheless, it still faces serious sustainability challenges. Innovative approaches to address these challenges have include strengthening of its ICT consultancy services, as well as signing MOUs with partner universities so as to improve working relations. For example,

---

17 So far, 18 countries have signed the Charter: Kenya, Senegal, Mauritania, Mali, Cote d'Ivoire, Tanzania, Mozambique, the Democratic Republic of Congo, Benin, Ghana, Guinea, Burkina Faso, Niger, South Sudan, Sudan, The Gambia, Guinea Bissau, and Nigeria (*www.avu.org*).

18 *www.avu.org*.

19 Which is divided into two options: electricity production and heat production.

20 I.e. food security information systems and networks, reporting food security information, availability assessment and analysis, baseline food security assessments, market assessment, nutrition status assessment and analysis and formulation and implementation of food security policies.

rather than relying on individual students fees, this burden is shifted to partner organisations, who will now enrol their students for specific courses (Easter and Ewins, 2010).

The Kwame Nkrumah University of Technology (KNUST)

Another good example is the Kwame Nkrumah University of Science and Technology (KNUST) that has expanded into ten regions by establishing strategic partnerships that enable them to outsource professors and other professionals from the private sector (Easter and Ewins, 2010: 14). KNUST was established following market research that found that there was a need for quality education that could be delivered through online methods. The fact that this university contracts local professors and professionals gives the courses a local perspective that is suited to their context and has been well received by students (ibid.: 20). KNUST is clearly an example of a programme that started as a home-grown solution, market oriented and using local capacities such as contracting local lecturers.

African Ministerial Council on the Environment (AMCEN)

Another cluster of continent-wide e-learning for sustainability initiatives was spear-headed by the African Ministerial Council on the Environment (AMCEN)[21], based on Decision 6 (AMCEN-12) that was later on reiterated by Decision 5 (AMCEN-13)[22]. This decision has sparked a Pan-African e-learning for environment network that cuts across different sectors of African society[23].

At the level of higher education, the Mainstreaming Environment and Sustainability in African Universities (MESA) has 80 universities from 30 African countries[24] participating in a wide range of activities such as developing materials on environment themes like green economy, climate change adaptation and ecosystems management (AMCEN, 2012). MESA has facilitated joint research inputs by seven universities in developing the World Water Day 2011 report, seminars on sustainability, curriculum sourcebooks on environment themes, a toolkit to support sustainable environmental practices in universities, Education for Sustainable Development in Africa (ESDA) in partnership with the United Nations University (UNU) and UN-HABITAT, through which a graduate curriculum had been developed in

21 For an overview of AMCEN and other institutions of the African Union and their roles in fostering sustainability in Africa, see generally, Takang and Njuki (2012). *African Union (AU)*. The Encyclopedia of Sustainability: Vol. 9. Afro-Eurasia: Assessing Sustainability (pp. 22-26).

22 Decision-6 (AMCEN-12) on Environmental Education and Technology-supported Learning was made at the 12th Session of AMCEN in 2008 held in Johannesburg and supplemented by Decision 5 at the 13th Session of AMCEN. The decision called among other things for countries to establish an African environmental e-learning network to share expertise, best practices and content and to identify a coordinating hub. See generally, *http://www.unep.org/roa/Amcen/ Amcen_Events/12th_Session_Amcen/docs/AMCEN-Ministerial-Segment-Report.pdf* (Last visited on 13 November, 2013).

23 During the 14th session of the African Ministerial Conference on the Environment held in Arusha, Tanzania from 10-14 September 2012, the progress made on implementation of environmental education and technology supported learning was discussed. A network had been established to facilitate e-learning for the environment and a number of sub-regional hubs were operational. However, not all countries were participating. See generally, *http://www.unep.org/ roa/amcen/Amcen_Events/13th_Session/Docs/14th%20 Session/K1281899.pdf* (Last visited on 13 November, 2013).

24 See generally *http://www.unep.org/training/programmes/mesa.asp* (Last accessed on 13 November, 2013).

urban environmental management; rural development, and natural resources management issues (AMCEN, 2012).

Outside the higher education sector, AMCEN Decision 6 prompted UNEP to organise a number of training workshops on the theme *e-learning in Practice* (eLIP) between 2008 and 2012. Another outgrowth of the AMCEN initiative is the Marketplace for Environmental Training and Online Resources (MENTOR)[25].

Like the AVU, financial sustainability has proved to be a major challenge[26]. This initiative was largely financed by the German bilateral organizations, and when GTZ, InWent and DED merged into GIZ, the support provided by InWent, which until then had been the capacity building arm, ended (AMCEN, 2012: 4).

Beyond these, short-term blended courses[27] offered by organisations to enable developing countries to better participate in international governance have sprouted across the African e-learning for sustainability landscape. For example, Climate and Development Knowledge Network (CDKN) offered a course to enable more effective participation in global climate change talks. As part of a three-year capacity building program, a three-day workshop was held in Doha prior to the start of COP 18 Climate Change Conference in 2012. This was followed by an eight-week e-learning course that was implemented in collaboration with Ricardo-AEA and the United Nations Institute for Training and Research (UNITAR)[28].

From the case studies presented here, one can summarise that e-learning for sustainability initiatives in Africa have predominantly targeted formal education and particularly the higher education sector. It is discernible that there are indeed home-grown initiatives such as KNUST, while AVU is externally driven and larger in scale, while MESA is continent wide, donor-supported and designed to infuse ESD into university education and to create public awareness.

In the following section, we use a simple set of criteria to assess the extent to which e-learning can foster sustainability in Africa.

An Assessment of e-learning for Sustainability in Africa

The *4As*: accessibility, affordability acceptability and adaptability (Tomaševski 2001) are globally accepted in the evaluation of the level to which the right to education is guaranteed. For this chapter, this set of criteria has been adapted to assess e-learning for sustainability in Africa.

---

25   MENTOR is hosted on UNEP website and contains guidelines, manuals, courses and modules, and a trainers' network for proper delivery of the courses. See generally *http://www.unep.org/mentor/about mentor.asp* (Last visited on 13 November, 2013).

26   The heavy reliance on donors who were not able to continue supporting the initiative had prevented activities taking place in 2012. This is indicative of the externally driven nature of the initiative. UNEP as interim coordinator of the network until steering committee is established. Since 2008, the networks met but in 2012, there was no funding.

27   Blended learning describes a situation where e-learning takes place in a mix of virtual and face-to-face environments.

28   See generally, *http://cdkn.org/project/delivering-training-and-capacity-building-for-negotiators/ ?loclang= en_gb* (last visited on 13 November, 2013).

Accessibility

The 2011/2012 ICT *access* and usage household and individual survey reports a rapid adoption of internet use in Africa via mobile phones, spurred by the diffusion of smart phones. In all countries other than Rwanda and Ethiopia, mobile phone ownership is now over 40 percent (Calandro *et al.,* 2012: 1). There is a definitive upwards trend in internet use especially through mobile phones on the African continent. In terms of accessibility though, it is observed that unequal access to the internet, mobile coverage and smart technology still pose serious limitations to e-learning for sustainability.

Access can be regarded from different perspectives. First, judging by poverty statistics, a large proportion of the population in most African countries is financially incapable of paying the fees that would enable them participate in e-learning programs. Even with OERs that are mostly free, the financial costs are contained either in the individual access to equipment or internet, which then pose a serious hindrance to many aspiring participants. Secondly, technological drawbacks limit the benefits that participants can derive from available courses. In a survey covering 413 practitioners involved in e-learning in Africa, it was established that 83 per cent of respondents used laptops and 71 percent use mobile phones daily for learning purposes (Isaacs *et al.,* 2013: 9), meaning that quite a significant proportion had access to the necessary equipment. However, 49 percent of the respondents had experienced problems with technology and infrastructure breakdowns (*ibid.*). Thirdly, living in a continent with very high levels of language diversity also poses challenges with regard to languages of instruction. The e-leaning Africa Report 2013 states that out of all the respondents, 40 percent produce local digital content but only 16 percent do it in indigenous African languages (Isaacs et al., 2013: 10).

It seems then, that in order to make e-learning for sustainability in Africa more accessible, it will have to be scaled up so as to reach the highest number of people with the technology available while at the same time being designed in a manner that allows for interactive, collaborative problem-solving. Furthermore, it should be cost-effective, easily manageable and employ technology that is not too high-tech that the cost of skilled manpower ends up being prohibitive.

Affordability

As far as affordability is concerned, several indicators apply: internet connection costs, the devices, software and maintenance and payment of instructors. The rate at which technological advances are being made in terms of hardware and software have an implication for the development and sustainability of e-learning programs in Africa, especially as most of them are still largely donor driven and hence contingent on funding and policies, both of which are in constant state of revision and change. Norris (2001) has argued that market liberalization of telecommunications services, coupled with a high-speed backbone, can permit African nations to 'leapfrog' stages of industrialization through new technology by investing in fully digitized telecommunications. Although Butcher (2013) noted that increasing affordability and availability of bandwidth have improved access, Barendse (2004: 65) asserts that liberalism has not been successful in addressing "the problem of affordability." A comparative study of ICT sectors in 17 African countries came to a similar conclusion: albeit partial or fully liberalised telecommunications sectors, access and affordability of services remains low compared with other major world regions (Calandro *et al.,* 2010). The authors assert that whereas increased affordability of devices is driving uptake, pricing of services constrains usage (ibid.).

Acceptability

Acceptability of e-learning initiatives can be appreciated at policy level as well as among the learners, teachers and other stakeholders. Acceptability would be the extent to which e-learning programmes are perceived to be useful and cost-effective enough for the end user. It is not enough to have the content and infrastructure set up, it has to be perceived to be a valuable addition to the learning experience of the end user.

For example, in order to increase the acceptability of its programme, KNUST has endeavoured to make the courses more relevant for the local context. They develop content that has direct relevance and resonance among the learners and include local, well known professionals in their roster of educators. This strategy is supposed to make their courses more attractive and demand driven.

Adaptability

For an e-learning programme to be considered adaptable, it should be flexible enough to cater for the changing individual learning needs of participants, while at the same time designed in a way that it can be adjusted to suit the changing local and regional environmental, economic and social context. On this front, Africa is not performing particularly well, mostly because most programmes such as the AVU were first conceptualized through a donor-driven, top-bottom approach or they have not been tailored and revised as required due to financial constraints. Butcher coins this elegantly when he states that current e-learning initiatives are largely using the innovation of OER to replicate content-heavy, top-down models of education that reinforce the student as being at the permanent receiving end of educational content (2013: 30). Hence, not only does the curriculum not show proof of an interdisciplinary approach in which the interconnected nature of sustainable development is explored, but the content also lacks adequate incorporation of local sustainability issues[29].

There has also been inadequate consideration of the pedagogic approach to e-learning programs. Emphasis has been placed on individual, self-paced, distance learning when developing e-learning programs (Abel and Long 2010: 8). However, greater effort should be made to develop solutions that will increase active participation and collaborative problem solving. The e-learning curriculum should be effective in tackling sustainability themes, complete with appropriate materials, and capable instructors. Furthermore, a multi-stakeholder approach and interactivity should become inbuilt aspects.

Even as African countries seek to develop a more dynamic e-learning environment, the explosion of the mobile telephone market has created new opportunities.

M-Learning – An Emerging Trend across Africa

The proliferation of mobile phones and other handheld devices such as Personal Display Assistants (PDA), iPods, MP3 and MP4 players, palmtop computers, and tablets has sparked a rapid and dynamic development of mobile learning (m-learning). In Africa, the growth has been fuelled by access to cheaper handsets, expansion of powerful 3G data transmission and the availability of prepaid mobile subscription packages (UNESCO 2011: 4). This mobility of

---

29   Some exceptions exist. The "Learning BioReg Meru", a pilot project of UNEP, the Governments of Germany and Kenya, as well as Universities from the two countries is said to focus on sustainability at the regional scale. See generally Ehlers 2007, E-Learning for Sustainable Development (SuDu).

technology, learner, and learning context expands the learning sphere, both within formal and informal learning areas (Hashemi *et al.* 2011). Furthermore, the development of mobile technology has coincided with the Web 2.0[30] revolution, both of which have increased interaction and collaboration. Through m-learning, it is possible to set up mobile 2.0 learning platforms, such as Moodle and Poodle and to use blogs, wikis, discussion boards, podcasts and other file sharing tools (Wang and Higgins, 2008). M-learning is rather an add-on and used in blended learning situations than as a stand-alone option.

M-learning increases access and inclusion by providing learning through technology that is affordable even for learners from poor resource backgrounds, especially through mobile phones (Hashemi, 2011: 2480). However, m-learning cannot replicate total learning experience due to physical device limitations. M-learning programs have good chances of sustainability, because learners contribute to the financial investment, as they are likely to buy mobile devices for their personal use anyway. In addition to this, mobile devices are less expensive to maintain, as compared to other technologies, such as networked computer systems. Therefore, m-learning provides a cost-effective solution for resource strapped institutions that are keen to keep abreast with technological advances. But even as mobile platforms are touted as the solution to Africa's lack of connectivity, due attention must be paid to the drawbacks. There are cost implications of accessing data via mobile internet. Furthermore, m-learning projects using off-the-shelf mobile devices could face problems of compatibility (Kukulska-Holme, 2007). For example, mobile devices cannot hold as many applications as computers. Furthermore, mobile devices have small screens, low resolution, slow processing and limited storage capacities. They also are characterised by low network speed and battery life (Zhang and Maesako, 2008). As a consequence, they cannot be used optimally to view or transfer large multi-media and data files. Especially in resource poor regions such as Africa, where the likelihood of lower quality devices and lack of electricity to regularly charge batteries is high, their use is hampered greatly.

With the recognition of how m-learning is revolutionising and blurring the line between formal and informal learning, UNESCO partnered with Nokia in 2012 on two themes: policy development and teacher capacity building. As an initial step, they commissioned working papers on policies related to mobile policies in 5 regions – Africa, Asia, Middle East, Europe and North America, and also how mobile phones are supporting teacher development. From their findings, mobile use is hindered through restricted use in formal school settings and is only happening through initiatives championed by individual teachers and organisations (UNESCO 2011: 8). Mobile, handheld devices, supported by wireless and mobile phone networks, can facilitate and support teaching and learning within and outside formal learning environments. To take advantage of this potential, it will take a change in culture and attitudes towards such devices, especially within the formal learning context.

Conclusion – A post-Cotonou Era?

The potential of e-learning to foster ESD in Africa is certainly enormous but a lot still needs to be done to harness this potential fully.

Firstly, there is need for professional development of teachers at all levels on integrating ESD into the curriculum and to enhance their abilities to use e-learning. In many countries,

---

30   Web 2.0 applications comprise in social media social as LinkedIn, Facebook, and other applications such as Dropbox and Google Maps.

professional development currently takes place through distance education, in-service training, and workshops. E-learning too can provide teachers with individual, self-paced training to enhance their ESD skills. Moreover, teachers at all levels should be trained to flexibly blend different teaching and learning methods. For instance, if teachers received digital materials which they can use via electronic equipment such as laptops and projectors, they should be capable of using them to supplement their teaching content in the formal curriculum. Through simulated exercises, they could help their students get closer to real life situations in which they have to make decisions on sustainability issues.

E-learning in Africa has had an institutional, formal approach, but maybe there are more effective ways to teach sustainability. A mix of formal, non-formal and informal learning could allow for continuous, practical learning which increases learners' interaction with the environment.

Beyond inclusion, there is also the important question of the quality of content provided through e-learning opportunities: To what extent can e-learning guarantee both? The pilots with mobiles, tablets, and other handheld devices point towards a trend where developers are constantly seeking ways to increase access to e-learning opportunities, but there is an urgent need to focus as much attention to ensure that these new technological developments correspond with a continuous questioning and refining of content towards curricula that are more stream-lined to impart sustainability values among the learners. Stakeholders in Africa need to find convergence points where traditional educational approaches meet technologically advanced digital learning, and where issues of quality are addressed while at the same time striving towards greater inclusion of African citizens in e-learning initiatives.

Another serious issue that must be addressed is that of the environmental impacts of the hardware. Billions of computers, laptops, PDAs, smart phones and similar devices with low durability could add to already huge existing dumps of e-waste across Africa. Additionally, due attention must be paid to the energy needs required to power these devices. Adopting low energy equipment, combined with the development of renewable energies and efforts to reduce energy costs could make e-learning more accessible, but also more environmental friendly (Abel and Long, 2010: 3).

Financial sustainability will go a long way to foster originality of content to suit local re-alities. As observed, e-learning initiatives that are donor funded and whose conceptual design did not originate in African countries have problems transitioning projects to the end users and are not required to prove sustainability to continue receiving funding. Hence in Africa, e-learning programs are still largely donor-driven and not in line with local realities. Consoli-dating financial resources locally, nationally or even regionally will be crucial to domesticat-ing e-learning for sustainability in Africa in the post-Cotonou era.

References

Abell, Thomas E. and Trey Long (2010): e-learning in Africa: Transforming education through enabling technologies. How a combination of technology innovations will drive new models for education in Africa. Accenture.

African Ministerial Conference on the Environment (AMCEN 2012), Report on implementa-tion of Decision-6 (AMCEN-12) and Decision-5 (AMCEN-13). Fourteenth session, Aru-sha, United Republic of Tanzania, 10-14 September 2012: AMCEN/14/INF/2.

Barendse, A. (2004). Innovative regulatory and policy initiatives at increasing ICT connectivity in South Africa. *Telematics and Informatics* 21 (1), 49-66.

Boix Mansilla, V., Gardner, H., and Miller, W. (2000). On Disciplinary Lenses and Interdisciplinary Work. In S. Wineburg and P. Grossman (Eds.), *Interdisciplinary Curriculum: Challenges to Implementation* (pp. 17-38). New York: Teacher College Press.

Boko, M., I. Niang, A. Nyong, C. Vogel, A. Githeko, M. Medany, B. Osman-Elasha, R. Tabo and P. Yanda, 2007: Africa. Climate Change 2007: Impacts, Adaptation and Vulnerability. Contribution of Working Group II to the Fourth Assessment Report of the Intergovernmental Panel on Climate Change, M.L. Parry, O.F. Canziani, J.P. Palutikof, P.J. van der Linden and C.E. Hanson, Eds., Cambridge University Press, Cambridge UK, 433-467.

Butcher Neil (2013). OERs and MOOCs: Old Wine in New Skins? In Isaacs S (ed) 2013. *The e-learning Africa Report*, ICWE: Germany.

Calandro, Enrico, Christoph Stork, and Alison Gillwald (2012). Internet Going Mobile. Internet access and usage in 11 African countries. Research ICT Africa Policy Brief 2, 2012.

Calandro, Enrico, Alison Gillwald, Mpho Moyo and Christoph Stork (2010). Comparative ICT Sector Performance Review 2009/2010. Towards Evidence-based ICT Policy and Regulation. Volume 3, Policy Paper 5, 2010.

Cantoni V., Cellario M. and Porta M. (2004). Perspectives and challenges in e-learning: Towards natural interaction paradigms. *Journal of Visual Languages and Computing*. 15, 333-345.

Carolan, M. (2008). The Multidimensionality of Environmental Problems: The GMO Controversy and the Limits of Scientific Materialism. *Environmental Values*, 17, 67-82.

Castells, M. (2002). The Internet Galaxy. Oxford University Press, Oxford.

Dreze, J., and Sen, A. (1995). India Economic Development and Social Opportunity. Delhi, India: Oxford University Press.

Easter, Stuart and Rory Ewins (2010). Funding for E-learning in Africa: A question of sustainability. Available at: *http://speedysnail.com/2010/Easter_and_Ewins_2010.pdf* (Last visited on 13, November, 2013)

Ehlers, Ulf-Daniel. E-learning for Sustainable Development (SuDe). Actes du colloque Initiatives 2005 [en ligne], Débat thématique 2, 2 mars 2007. Disponible sur Internet: *http://www.initia tives.refer.org/Initiatives-2005/document.php?id=168.*

Engelbrecht E. (2003). A look at e-learning models: investigating their value for developing an e-learning strategy. *Progressio*. 25(2), 38-47.

Fuchs, Christian and Eva Horak (2006). Africa and the digital divide. *Telematics and Informatics* 25 (2008),99-116.

Hashemi, Masoud; Masoud Azizinezhad; Vahid Najafi and Ali Jamali Nesari (2011). What is Mobile Learning? Challenges and Capabilities. 2nd World Conference on Psychology, Counselling and Guidance – 2011. *Procedia – Social and Behavioral Sciences* 30, pp. 2477-2481.

Hesselink, Frits; Pert Paul van Kempen and Arjen Wals (2002). ESDebate: International Debate on Education for Sustainable Development. Gland: IUCN Commission on Education and Communication (CEC).

Hilty, Lorenz M., Aebischer, Bernard, Andersson, Göran and Lohmann, Wolfgang. ICT4S 2013. Proceedings of the First International Conference on Information and Communication Technologies for Sustainability, ETH Zurich, February 14-16, 2013. ETH Zurich, University of Zurich and Empa, Swiss Federal Laboratories for Materials Science and Technology (2013). *http://dx.doi.org/10.3929/ethz-a-007337628*

Holmes B. and Gardner J. (2006). E-learning: Concepts and Practice. Sage, London, U.K.

Huckle, John and Stephen Sterling (1996). Education for Sustainability. London: Earthscan Publication Ltd.

Isaacs S, Hollow D, Akoh B, and Harper-Merrett T. (2013). Findings from the e-learning Africa Survey 2013., in Isaacs S. (ed) 2013. *The e-learning Africa Report*, ICWE: Germany

Kop Rita, Hélène Fournier and John Sui Fai Mak (2011). A Pedagogy of Abundance or a Pedagogy to Support Human Beings? Participant support on Massive Open Online Courses. International Review of Research in Open and Distance Learning, 12 Special Issue – Emergent Learning, Connections, Design for Learning, 7, pp. 74-93.

Kukulska-Hulme, Agnes (2007). Mobile usability in educational contexts. What have we learnt? *International Review of Research in Open and Distance Learning* 8 (2), 1-16.

McKeown Rosalyn, Charles A. Hopkins, Regina Rizzi and Marianne Chrystalbrid (2006). Education for Sustainable Development in Action Learning and Training Tools N°1. UNESCO, Paris.

Munson, K. G. (1997). Barriers to ecology and sustainability education in the U.S. public schools. *Contemporary Education*, 18(3), 174-76.

Neumayer, E (2003). Weak versus strong sustainability: exploring the limits of two opposing paradigms, Elgar, London.

Norris, Pippa (2001). Digital Divide. Civic Engagement, Information Poverty, and the Internet Worldwide. Cambridge University Press, New York.

Pfeffermann, Guy (2013). New Models of Management Education for Africa. In Isaacs S. (ed) 2013. *The e-learning Africa Report*, ICWE: Germany.

Rodriguez, Oswaldo (2013). The Concept of openness behind c and x-MOOCS (Massive Open Online Courses). Open Praxis, Vol. 5 Issue 1, January-March 2013, pp. 67-73.

Schultz, T. W. (1961). Investment in Human Capital. *American Economic Review*, 51, 1-17.

Sitarz, Daniel (Ed.). (1993). Agenda 21: The Earth Summit strategy to save our planet. Nova Publishing Co.

Sterling, Stephen (2012). Sustainability Education: Perspectives and Practice across Higher Education. CRC Press

Takang, J. Manyitabot, and Njuki, Caroline Muthoni. (2012). African Union (AU). In Louis Kotzé and Stephen Morse (Eds.), *The Encyclopedia of Sustainability*: Vol. 9. Afro-Eurasia: Assessing Sustainability (pp. 22-26). Great Barrington, MA: Berkshire Publishing.

Takang, J. Manyitabot. (2012a). Water use and rights (Africa). In Louis Kotzé and Stephen Morse (Eds.), *The Encyclopedia of Sustainability*: Vol. 9. Afro-Eurasia: Assessing Sustainability (pp. 365-372). Great Barrington, MA: Berkshire Publishing.

Takang, J. Manyitabot. (2012b). International Conflict Resolution. In Louis Kotzé and Stephen Morse (Eds.), *The Encyclopedia of Sustainability*: Vol. 9. Afro-Eurasia: Assessing Sustainability (pp. 168-172). Great Barrington, MA: Berkshire Publishing.

The William and Flora Hewlett Foundation (2010) Open Educational Resources. Available from: *http://www.hewlett.org/programs/education/open-educational-resources* (Accessed February, 28, 2014).

Tilbury, D. (1995). Environmental education for sustainability: Defining the new focus of environmental education in the 1990's. *Environmental Education Research*, 1(2), 195-212.

Tomaševski, Katarina (2001). Human rights obligations: making education available, accessible, acceptable and adaptable. Right to Education Primers No. 3. Gothenburg: Novum Grafiska AB.

United Nation (2013). The Millennium Development Goals Report 2013. New York: United Nations.

UNDP (2005). United Nations Human Development Report (UNHDR), 2005, New York.

UNEP (2007). Global Environment Outlook 4: Environment for Development. United Nations Environment Programme (UNEP).

UNESCO (2011). Mobile Learning Week Report, 12-16 December 2011, UNESCO HQ, Paris. Available at: *http://www.unesco.org/new/fileadmin/MULTIMEDIA/HQ/ED/ICT/ pdf/UNESCO% 20MLW%20report%20final%2019jan.pdf* (last visited on 14 November, 2012).

Van Dijk, J., 2006. The Network Society. Social Aspects of New Media, second ed. SAGE, London.

Wang, Shudong and Michael Higgins (2008). Mobile 2.0 Leads to a Transformation in M-learning. In: *Proceedings of ICHL*. Berlin: Springer, pp. 225-237.

Weigel, V. B. (2002). Deep learning for a digital age: technology's untapped potential to enrich higher education. San Francisco: Jossey-Bass.

Wilson, E.J. (2006). The Information Revolution and Developing Countries. MIT Press, Cambridge, MA.

Wood, Gillen (2012). "Sustainability Studies: A Systems Literacy Approach", in Tom Theis and Jonathan Tomkin (Eds.), *Sustainability: A Comprehensive Foundation Collection,* pp. 504-508. Available at: *http://cnx.org/content/m41039/1.5/.*

World Commission on Environment and Development (WCED). (1987). *Our Common Future.* Oxford: Oxford University Press.

World Summit on the Information Society (WSIS), 2003. Declaration of principles. Building the Information Society: A Global Challenge in the New Millennium. Available from: *http://www.itu.int/wsis/docs/geneva/official/dop.html* (Last visited on 13 November, 2013).

Zhang, Hai and Takanori Maesako (2008). A Theoretical Framework of Ecosystem of Learner Development for Designing a Practical Ubiquitous Learning Environment. In: Leung, Elvis Wai Chung (Ed.): *Advances in Blended Learning.* Second Workshop on Blended Learning, WBL 2008, Jinhua, China, August 20-22, 2008. Revised Selected Papers. Berlin: Springer, pp. 83-92.

Sustainability in an educational institution: analysing the transition to paperless e-processes, an Indian case

Prakash Rao[1], Yogesh Patil[2], Manisha Ketkar[1],
Viraja Bhat[1] and Shilpa Kulkarni[1]

Abstract

In recent times, environment and sustainability have caught the attention of various sections of society as these issues are threatening the very fabric of global systems and processes. The recent Rio +20 summit in 2012 has reiterated the role played by education in building sustainable development recommending amongst other issues newer methods of pedagogy and innovative processes in institutions of higher education.

The present chapter focuses on a new approach towards a sustainability based academic process through the development of an online system using Information Technology tools at an Indian Business school. This includes tools like Faculty Information Systems, online testing, online student performance appraisal and feedback systems etc.which have been designed to create improved academic efficiency, higher stakeholder satisfaction levels along with potential environmental co benefits like reduced paper usage. The need for incorporating sustainability as a core element of higher education has been growing over the years in Indian Universities and the study is a step towards providing solutions by using IT systems. Key lessons learnt from the study revealed that replicability of this approach in other institutions is dependent on several factors like stakeholder satisfaction, infrastructure, skilled workforce, top management support and process standardisation. Better efficiency and use of smarter IT based tools and indicators as a means towards achieving sustainability in management education will have significant implications for academic process development at Universities as part of the UN decade of Education for Sustainable Development (2005-2014).

Introduction

Education has played a key part in driving the paradigm of growth across post independence India. Since 1947, Independent India has had a major thrust towards industrialisation and agriculture, and building the country's granaries to fulfil the needs of its citizens was a key priority. Building a team of qualified professionals across the rest of the country to serve the needs of the industrialized world was a key priority of the policy makers. India's English literate population is a key to the competitiveness (Porter, 1990). As an emerging economic power house the human capital in India is expected to make significant strides shaping the economic growth paradigm. The importance of education was recognized by fore fathers of

---

1   Dept. of Energy and Environment, Symbiosis Institute of International Business, Symbiosis International University, Pune, India.

2   Head – Research and Publications, Symbiosis Institute of Research and Innovation, Symbiosis International University, Pune, India.

nation even in pre-independence time with several national stalwarts like Mahatma Gandhi, Mahatma Phule, Sarvapalli S Radhakrishnan, Bharatratna D.K. Karve who have paved the way for bringing a new dimension to the Indian education system. Post-independence India has not only moved on the path of education and literacy but also brought in laws relevant to education to protect the right to learn for each Indian.

As a major economic growth centre India aims to be the skill capital of world (NSDCL, 2011). Post-independence India has witnessed an above average growth in the number of higher educational institutions vis-à-vis its population. India had just about 20 Universities and 500 colleges at the time of independence in 1947. As a part of the Nation building activities several important initiatives were taken up aimed at increasing the knowledge base of the country through the establishment of specialised Institutions like the Indian Institutes of Management (IIMs) and the Indian Institutes of technology (IITs). Simultaneously, other educational institutions also started developing their own centres aimed at imparting quality training and knowledge based education.Today, these numbers have grown exponentially. In the higher education sector India today has a total of 610 universities. 43 central universities, 299 state universities, 140 private Universities, 128 deemed universities and 5 institutions established through state legislation, 30 Institutions of national importance.The total government spending on education is approx. 3.8% of its GDP on education and FDI inflows in the education sector during May 2012 stood at $31.22 mn (Deloitte, 2012).

In general, education which was considered a service in the earlier times is now seen by some as a business which serves the society at large for the development of an individual. Hence the effectiveness of the education depends not only on the concept of teaching and learning but also governance models and regulatory mechanisms. Lot of emphasis is given by the governments for the basic education and over the years the scenario of primary and secondary education has improved considerably. For the social and economic development of the country, higher education is of utmost important. According to a recent report by the Economist Intelligence Unit (2008), technological changes will effectively change the skillsets of the future workforce, as well as have a positive impact on campus administrative processes and systems.

Higher Level education in India

The institutional framework consists of Universities established by an Act of Parliament (Central Universities) or of a State Legislature (State Universities), Deemed Universities (institutions which have been accorded the status of a university with authority to award their own degrees through central government notification), Institutes of National Importance (prestigious institutions awarded the said Status by Parliament), Institutions established State Legislative Act and colleges affiliated to the University (both government-aided and unaided).

There are three primary stages of qualification within the higher education system

1. Graduation level;
2. Post-graduation level;
3. Doctoral degree.

Besides these there are other qualifications like Diploma and certificate courses which can be added on after graduation or post-graduation. Some diploma courses are offered as skill based training especially in engineering sector.

Structure of Indian Education sector

The regulatory body for the higher education in India is University Grants Commission (UGC) and for professional courses it is All India Council for Technical Education (AICTE). There are also private universities which are under state law and governed as per State law. The institutes which have run professional courses are affiliated to the AICTE, while simultaneously many professional courses which are under autonomous institutes and have deemed university status are not affiliated to AICTE (Twelfth Five year plan of the Government of India percentual progression can be seen in figure 1).

Figure 1: Student Enrolment in Higher Education (million) in India.

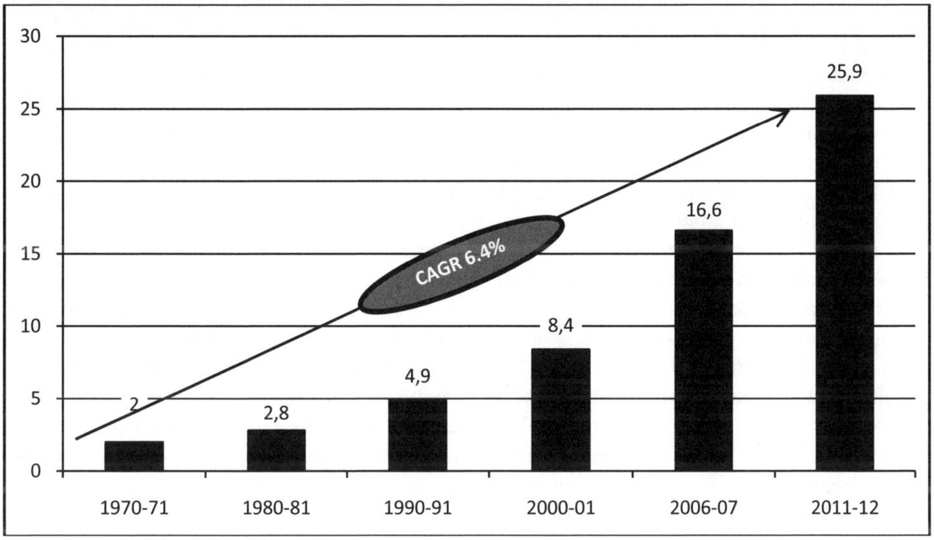

Source: Higher Education in India: Twelfth five year plan (2013-2017) and beyond, FICCI report.

Societal angle

With the adoption of the Rio +5 declaration in 2002, various approaches were developed globally by academic institutions in designing and operationalizing sustainable development activities at Universities through either legislations, regulatory frameworks, policies or voluntary standards.Around this period the Luneberg Declaration (2001) was adopted which called for promoting core values of sustainable development in all forms of higher education as an integral part. This was further developed in 2005 with the commencement of the UN Decade of education for sustainable development. Incorporating the principles of sustainable development for higher education activities was one of the major objective of the Copernicus guidelines for developing sustainability in higher education systems (Copernicus, 2010). One of the key features of these guidelines included cost-effectiveness of higher education systems and governance process management. The Washington based Aspen Institute has created a systematic ranking system of Business Schools around the world which have taken on sustainability initiatives (Godemann et al., 2011).It is also true that while more than 600 Universities worldwide have committed themselves to various aspects of promoting sustainable development, very few have been able to successfully integrate the principles of sustainable development into practice due to a combination of reasons, varying from lack of institutional interest, to limited resources or staff involvement (Leal Filho, 2011; De Kraker et al., 2007). In the Asia Pacific, regional attempts have been made to get a preliminary understanding of trends in sustainability in business education (Naeem and Neal, 2012)

Post liberalization after 1991 India has seen many changes on the path of economic growth. This economic growth has been compounded by the educational reforms in higher education. Management education has seen growth in leaps and bounds especially after the change towards liberalization, privatization and globalization. India being the part of global linkage in the aftermath of the WTO agreement is becoming a technology driven society. Presently more than 3000 management colleges exists across the India and many among them like the Indian Institute of Management (IIM) are counted among the best. India's entry in the globalization process also acted like a catalyst in this management education boom. Because of the increasing number of global and Indian multinationals, trained management graduates are in demand. In response to the growing demand, many private sector initiatives have also seen development of niche based higher education.

The manager being the key strategic game changer, it is essential to invest in the concept of holistic or sustainable business concept into management education (Blackburn, 2009). Educational institutes being the center for learning contemporary management practices, it becomes more than necessary to understand the sustainable ways for managing the business. Awareness and education are the crucial components for bringing the change for any business including evolving new governance paradigms in administrative mechanisms using Information and Communication Technologies (ICT) based tools.The concept of implementing sustainability through ICT based technologies and other operational mechanisms has evolved considerably in recent times (Carbon Trust, 2006). These are evident through studies on the environmental impacts of campuses and educational institutions (Leal Filho, 2009; Pandey et al., 2011; Roy et al., 2008). Several universities have also initiated research and action to assess their GHG inventory (University of Toronto, 2010) and tracking their carbon emissions profile (Sprangers, 2011). The importance of sustainability and its linkages with e-learning is therefore is now being seen as a major aspect of sustainability measurements and indices.

Role of ICT in education

Across the world, in the last decade of the 20th century significant changes have been seen in ICT as a transformational technology which had a profound societal effect. Through ICT our society is changing from an industrial society to information or knowledge based society. While in the industrialized society the main focus of education was to contribute to the development of factual and procedural knowledge, in the information or knowledge driven world, the development and management of knowledge is increasingly considered a key factor of progress.This change undoubtedly has implications for our education systems and has led the path towards betterment (Mihane and Badivuku-Pantina, 2010).The effective and innovative integration of technology in education is considered the passport to the third millennium for the overwhelming majority of citizens across the world (Srivastava, 2010). A framework of knowledge, skills and attitudes coupled with successful task performance and problem solving with regard to sustainability will therefore be necessary for academic programme development (Barth et al., 2007).

Changes have been witnessed in every sector of the growth oriented industry at uneven pace. The education sector has also been part of this change and in the era of technology we live in, the integration of the information technology tools is an inevitable component in the education sector administration. Tremendous growth in the education sector poses a challenge for administration of the various activities in an educational set up. The processes followed in the governance and administration needs to be effective and efficient in terms of communication and cost.

Universities, both in the developed world and the developing countries face a new challenge of equipping students with adequate education in their field of study, and also arm them with the skills and knowledge to leverage technology effectively in the workplace. In a developing country like India after Independence, there has been a tremendous growth in higher education sector.In India, the students lack access to higher education due to factors of availability and affordability. In the Twelfth Five year plan of the Government of India, there is increased focus on not only the number of Universities but also on quality of education through setting up of several universities as centers of excellence and Innovation based universities. These new initiatives have tried to cater to the needs of the students but the increase in quantity is still to be quantified in terms of quality and relevant output. Recent studies (Bandalaria, 2007) linking education and governance and ICT seems to suggest that newer methods of administrative function can go a long way in improving the quality of and delivery of academic programmes in higher education in emerging economies. Technological solutions and innovative services could therefore be the key to solving complex administrative and governance issues.The scale of disruption and discontinuity faced in the existing educational processes may result in the role of technology in education as a highly sensitive and emotional issue.

Educational institutions generally tend to use ICT in the following functional areas – General Administration, Payroll & Financial Accounting, student data administration, Personnel record maintenance, Inventory management and Library systems.

A high penetration of technology effectively can help in building a system which is students centric and transparent, self-servicing, efficient and low cost will improve the state of education (Mooij, 2008). The big question to be dealt with is where should the institutions start their journey? It is a challenge to make all the processes and resources open for people through an online system from visualisation to the end user.

Methodology

The present study deals with a unique effort to develop a reliable, efficient and cost effective process to enable key stakeholders of the education system to benefit and realize improved efficiency in the delivery of higher education. The authors have been responsible in designing the techniques for the progression of the system for replicability into other institutions of higher learning. We used the sustainability matrix of social, economic and environmental dimensions (Prahalad, 2006) to examine the transition to introducing ICT based systems at the Institute. This included assessment of stakeholder satisfaction levels (social), economic benefits (cost savings) and resultant environmental co benefits (reduced paper usage). A qualitative cum quantitative approach was used to record the changes in the academic processes at the Institute through a semi structured response survey with students as main stakeholders apart from quantifying the various academic processes (admissions, performance appraisal system, feedback mechanisms, examination processes etc.). The comparison with the earlier administrative process was a key factor in designing new and innovative processes.

Student communities from three different Post graduate streams and batches (2011-13, 2012-14 and 2013-15) were considered for the survey to gauge satisfaction levels of online systems.

The Symbiosis approach

Symbiosis International University was the initiative of Padma Bhushan Dr.S.B.Mujumdar a renowned educationist who initially focused on bringing harmony and better education for foreign nationals studying in Pune, India. His visionary approach towards inclusive and quality education helped create the Symbiosis International University (SIU).Based out of Pune, SIU as an academic institution initially began its journey through establishing an institution focusing mainly on supporting foreign students, teaching them the English language and then diversifying into education, including management education, as the means to promote international understanding. Over the past few decades the University has grown as one of India's top Academic Institutions catering to the needs of graduate, post graduate and doctoral students in varied disciplines ranging from management science to law, media,communication and design, engineering, technology, health and biomedical sciences, computer studies, humanities and social sciences etc.

ICT based Processes and systems in a higher education Institute – the SIIB approach

The Symbiosis Institute of International Business (SIIB), Pune is a premier Indian Management Training institute and implements world class teaching and administrative practices. SIIB, a constituent of Symbiosis International University, Pune was established in 1992 with a view to provide management education in International Business following the economic liberalisation efforts in 1991. Over the past twenty years, the Institute has focused on education, research and training in the field of International Business and other niche based programmes to equip aspiring global managers with managerial, economic, legal and relevant technical competence. The Institute conducts full time Post Graduate Management programmes for students in the field of International Business, Agri Business management and Energy and Environment apart from conducting customised training programmes in these areas.

The problem

Till 2008, most of the Institute's academic and administrative systems were handled manually including processes that involved timetable management, attendance management, attendance requisition, display of marks to students, submission of student projects, feedback, etc.This often led to series of administrative delays and lack of clarity over providing students with timely data and information related to their academic assignments. These are detailed below.

Students joining the Institute after the formal admission process were also distributed hard copies of the academic manual as also various class assignments. The process of recording student attendance and marks using softwares like MS Excel took almost a couple of weeks to publish. As a result, students and faculty members found it difficult to keep track of attendance on particular days of the earlier months. There were also difficulties in filing and reviewing attendance requisitions. At the end of a semester, there used to be several queries related to attendance/marks and student evaluation where students often were not able to meet certain academic norms.The task of data retrieval from files involved long hours and often remained incomplete leading to student dissatisfaction with academic departments in terms of timely information delivery related to their academic performances and other processes. Students also had to take paper copies of their assignments for submission which involved extra costs (paper purchase) and additionally the Institute had to find adequate storage space each year to keep these documents. The present approach using existing manual systems also involved logging in data related to compilation of faculty hours and faculty feedback which involved considerable time and effort coupled with use of paper. Updating the academic time table on a regular basis and intimation of the changes in the evening or on weekends/holidays was often a tedious effort and often led to inefficiency in providing timely information and confusion amongst various stakeholders like student community, faculty, administrative staff etc.

The need

Based on the real time problems being faced by stakeholders in the Institute for delivering efficient, reliable and timely academic information there was a critical need felt by the Institute administration for developing an approach to improve the efficiency of the administrative process. Given that Information Technology was a key driver of improved efficiency of administrative processes, the Institute developed a system which would be compatible to not only student community but also for other stakeholders.

System and process Improvements

SIIB saw the improvement opportunities and decided to replace the manual functions with online systems. The process involved not only streamlining of systems but also resulted in higher efficiency of processes, improved satisfaction levels amongst stakeholders and environmental cobenefits.

Exams/assignments, project submission

This was done in a phased manner starting with administering online tests for students. The process helped in generating different sets of question papers for each student in the same examination as different questions were selected through a random process. The system brought about efficiency and established better control in conduct of the examination. The result of the tests were instant with an added facility for the students to seek clarity on their answers. A common Information folder was created on the central server (accessible only in the teaching block) wherein information related to academic manuals, teaching presentations of faculty, technical reports, research documents etc. were made available rather than distribution of hard copies. Similarly, students also provided their class assignments/project submissions through the online process instead of a manual submission which was the earlier practice. The resultant changes benefited the Institute on account of reduced storage space and administrative difficulties of maintaining large volumes of the records.

Attendance systems

One of the innovative methods to record class attendance was initiated through the online process. This involved recording of student attendance in the classroom making it a very transparent system.The process resulted in tracking and updating of student attendance records and faculty teaching hours in real time. Absence from class by students on account of Institute related duty, was earlier resolved by respective faculty incharge through manual requisitions for those classes. Matching requisitions with missed classes and then updating attendance used to be a herculean task. This was solved on-line by displaying the details of missed classes to students and routing their requisitions to faculty-in-charge for review and approval. The attendance was updated automatically based on acceptance or rejections of the requisitions by the faculty. The online attendance system helped retrieve archival and manual records easily at the end of a semester particularly in cases where students fell short of attendance norms to appear for final examination or meeting up some of the internal norms.

Online marks system

As a part of the transition process, students marks were declared through the online process, giving students the opportunity to review marks and raise queries, if any.This also helped the students to monitor their academic performance on a regular basis and for the administration in maintaining up to date records. Additionally, faculty were also able to analyse students' performance and modify teaching and pedagogical methods as necessary.

Other Online systems

Feedback from students on faculty, peers, workshops, etc.,was also driven through the online process.The major advantage in doing this was to get timely feedback reports from students The timely feedback process helped improve teaching methods and processes further. An online calendar for displaying timetables/class schedules was also developed through a colour coding scheme and shared with respective students and faculty members. Dissemination of any changes to the schedules then became easier for students to track. For faculty members, an online faculty information system was developed which provided each member a facility to record and update their professional work profile including data on administrative responsibilities, research publications, paper presentations, training programs etc.This was also integrated with other online processes. The access to information therefore was streamlined and became more efficient and could be used for fulfilling academic compliances and other for other administrative needs like annual staff and faculty appraisals etc.

Table 1:   Cost Benefit Analysis of the Online academic administration process at SIIB (2011-12)

Sr.No	Functions/Processes	Paper sheets (A4) saved in Nos	Paper sheets (A4) saved in Kgs
1	Attendance Records	6060	30.3
2	Administration of records	162	0.81
3	Attendance requisitions	1600	8.0
4	Teaching notes / case studies in soft copies	8080	40.4
5	Assignment submission in soft copies + on-line tests	40400	202
6	Publication of marks	60600	303

Sr.No	Functions/Processes	Paper sheets (A4) saved in Nos	Paper sheets (A4) saved in Kgs
7	Grievances about marks and its feedback	400	2.0
8	Feedback	11110	55.5
9	Compilation and publication of feedback/year	404	2.02
10	Academic manual	6000	30
11	Time table	1680	8.4
12	Faculty information system	900	4.5
13	Admission process	40000	200
Total papers/sheets saved per year		**177396**	**886.98**
1	Total savings on paper per year in INR*		42575
2	Total savings on photocopying / printing per year in INR		135796
3	Total Savings on staff salary per year in INR		180000
Total savings per year in INR			**358371**

* Cost of a paper in INR 0.24

Table 1 represents the various processes used for transition from a manual system to ICT based processes. The table is an explanation of the economic and environmental benefits using paper and time as key indicators. In the higher education system especially in the institutions where the students are part of residential programs for a period of two years, two areas viz. knowledge administration and information administration are very crucial. The knowledge administration which gets the continuous focus by means of teaching and learning is part of the knowledge cycle with knowledge acquisition, knowledge assimilation and development and evaluation. In the higher education institutions, the information administration is equally important to cater to the high level of effectiveness and efficiency. The information administration mainly covers general and day to day operational activities. The information administration cycle includes the major components namely student administration, staff administration and general administration. For sustenance and decision making in Indian Higher educational institutions, there is a need to adapt to the different methods of administration supported by technology. The major challenge faced in the adoption is the implementation due to the complexity of the administrative tasks while dealing with the student administration (Petrides and Nodine, 2003).

These online systems have made the processes convenient, consistent, and comprehensible, and provide 24 X 7 accesses to information and communication. This has been made possible by timely two way communication with the help of a dedicated web page for individual stakeholder. Additionally, the ICT processes saved money and efforts.In today's technology driven educational institutions, the decision making by the stakeholders depends entirely on the real time communication. Unlike the brick and mortar model of education, notice boards have become redundant leading to saving paper and resultant carbon footprint and have made stakeholders effective by providing them the right information at the right time for right decision making.

Sustainability due to implementation of online process

The study compared past and existing academic processes and the benefits derived in terms of conversions to an online based system across social, economic dimensions and to some extent environmental dimensions in the journey towards achieving sustainability. The study focused largely on the implementation mode of various academic processes and its potential cost in terms of sustainability as an issue. Over the past few years educational institutions in emerging economies have also started to evaluate criteria like quality, service and environmental consciousness as a key requirements of providing holistic education practices (Cloquell – Ballester et al., 2008). The use of such ICT based systems has also led to resource optimization in terms of use of energy, paper, waste as well as shorter time and efficiency of academic student related operations (Eneroth, 2000; Garrison, 2000). The importance of delivery support services and management and administrative functions through ICT based application has also led to effective and faster administrative services to students and cost reduction (Bandalaria, 2007). At SIIB, the model focused mainly on operational improvements keeping in mind students as the key stakeholder. In terms of gauging the sustainability aspects of this transition to an online and electronic system the social, environmental and economic benefits were examined.

Economic Benefits

The online processing system generated a systematic process flow and hence the institute was able to save INR Rupees 358,371per year (Table 1). The systematic flow not only helped in process standardization but also helped to reduce human errors and improve operational efficiency for various processes.This intangible benefit has helped to increase reliability, and making systems much easier to operate. Although, the monetary benefits achieved year on year might seem relatively less, in the long run this is likely to have significant implications on savings.

Social Benefits

The social dimension of the model dealt with understanding the overall benefits and usefulness of a new online system created for access by students. An online student survey was conducted through the satisfaction levels of the new system as against the earlier manual system. The online survey questionnaire was circulated across 600 post graduate students across three different batches for the years (2011-13, 2012-14,and 2013-15). Responses were received from approximately 21 % of the students (n=127) from three different Post graduate full time programmes. The use of ICT based instruments and process like online testing, mobile telephone were key factors in coordinating academic activities between staff coordinators, faculty and students.

The majority of the responses (92 %) were from current batch students while 8% of the respondents were from the batch just passed out (Figure 2).

Figure 2: Student Batch wise response rate for assessing satisfaction levels of introducing online processes in academic system.

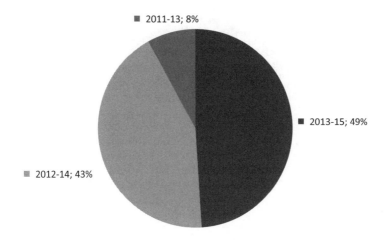

Figure 3: Programme wise response for assessing satisfaction levels of introducing online processes in academic system.

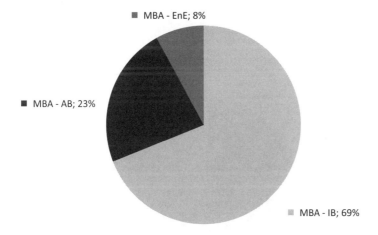

The bulk of the respondents (69 %) were from the flagship program MBA – International Business (IB) (Figure 3) where student strength was also the highest indicating that the eprocess system implementation was also reflective of the larger batch strength of the IB programme.

Figure 4: Satisfaction levels of respondents for introducing online processes in academic system.

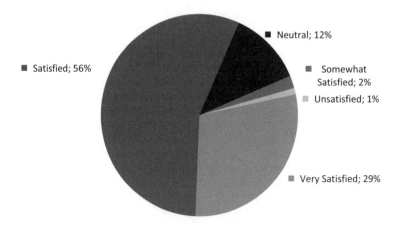

The survey also tested the efficacy of the quality and quantum of information provided to students. 85% of the respondents (Figure 4) agreed that online systems help in disseminating much more information quickly and easily. Online systems make compilation easy and eliminate human errors thereby increasing information consistency and ensuring that updated records are always available through the system. 81% respondents were happy with accuracy provided by online systems (Figure 5).

Figure 5: Satisfaction levels for accuracy and efficiency of academic information provided through online processes.

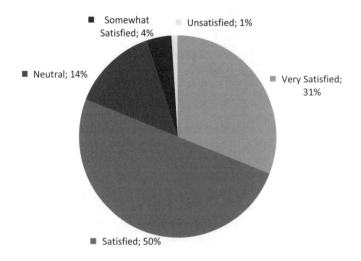

The introduction of the online systems helped collate and analyse data very quickly and made real time data available. Access to such data was also convenient at any hour of the day (Figure 6).

Figure 6: Satisfaction levels for timeliness of academic Information provided through on line processes.

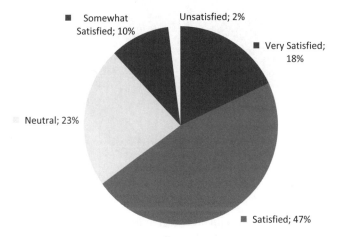

In an age of information overdrive, the use of technology and its applications has proven to be a boon of the younger generation.Online systems therefore are a preferred option. 89% respondents experienced ease in use of online system application process (Fig 7). The overall satisfaction levels amongst the student community in terms of using an electronic online system for their academic pursuits and administrative functions was as high as 89 % (Fig 8).

Figure 7: Satisfaction levels for use of online system application process.

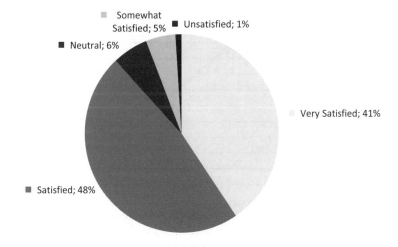

Figure 8: Overall experience of respondents for introduction of online processes.

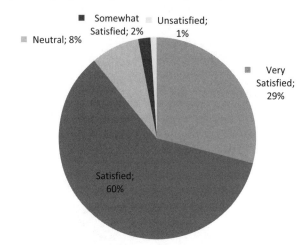

Marcuse (1998) observes, "Sustainability as a goal in itself, if we are to take the term's ordinary meaning, is the preservation of the status quo. It would, taken literally, involve making only those changes that are required to maintain that status."The transition made in the administrative process to an online system is a measure to maintain the status quo in a responsible, socially acceptable and economically viable manner.

The survey essentially queried the student respondent to outline the reasons for usage of the online system. The questionnaire was reliable enough to be analyzed further (Cronbagh Alpha: 0.789). Most of the respondents (n=127) utilized the online system for user friendliness rather than timely and accurate information (Figure 9).

Figure 9: Usage of the online administrative e process amongst student community.

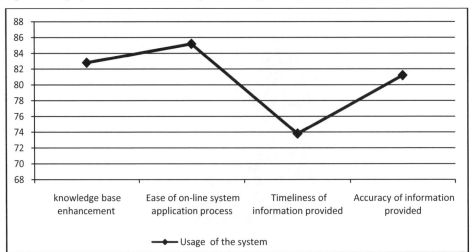

A comprehensive literature review by Murphy (2012) outlines a framework relevant for policy development and assessment focusing on four key dimensions: equity, awareness, participation, and social cohesion. Societal sustainability of the business can be structured on the basis of understanding and assimilation. Using an ICT based system is one of the examples where it can be seen that efficiency and transparency for providing information is critical. The ICT based system is also effective in making the end user or stakeholder aware of the recent administrative changes with timely and accurate information. These factors could be an indication towards building social cohesion in sustainability practices at campuses. As a consequence the new system is more pro stakeholder and invites active participation and feedback from the end customer.

If one fails to advance a system in which procedural aspects are taken into account (effective participation of social stakeholders), then one can expect few incentives to embrace goals and concerns that run counter to leaders' framing of the issue (CasulaVifell and Soneryd, 2012). Power struggles and limited access to information create obstacles in participation (Hallström and Boström, 2010).

Environmental co-benefits

The introduction of online systems as a function of academic administrative processes also considered environmental sustainability as a key co benefit for improved governance and development of organisation culture. The concept of a holistic sustainability initiative across campuses has not only been seen from an economic or social perspective but also in terms calculating the likely environmental footprint of an educational institution through operational and sustainable development activities (Leal Filho 2011; Sprangers, 2011). While this may be biased towards mapping various indicators of environmental sustainability through assessment of consumption of energy, resources and waste disposal, very few campuses have perhaps studied the transition to electronic based administrative functions as a measure of sustainability. During the present study the primary indicator considered for the environmental footprint aspect was the use of paper in academic administrative processes.

Before the transition to an online process, the manual system of recording data related to administrative processes entailed considerable use of paper during the initial phase of the Institute 's academic programmes i.e. 1992-2009. Consequently the progression to an online system from 2010 led to the savings of approximately 177396 (886.98 kgs) sheets of paper over one full academic year session (Table 1). The environmental implications of this savings in terms of the future growth of the Institute's academic programmes is significant in lowering the Institute's footprint and improving its sustainability levels. Assuming that nearly 90 % of the future administrative system will be converted to an online based system, this would mean resource conservation of nearly 22,012 kgs of paper saved, approx. 81646 kgs of wood saved for trees or the equivalent of about 2700 trees planted each year and about 100 tonnes of carbon dioxide emissions saved. In the long run the implications of resource conservation and sustainability will become a major factor in driving sustainability across campuses.

Limitations and replicability of the study

The study is limited to one management institute only with strength of approximately four hundred students and about fifty academic staff. It did not consider the potential impact of introducing online academic systems and process on the university as a whole. From a social assessment point of view the satisfaction levels of all the stakeholders could not be measured as some of the stakeholders (e.g. alumni, past employees, visiting faculty etc.) were not considered due to lack of data and inaccessibility of personnel. In terms of environmental impact the study only considered paper as a key indicator in terms of savings made from conversion to an online system.

We also advocate the replicability of this approach at other academic institutions as well. This will however be dependent on a number of criteria and factors related to technology, governance and human resources. The cost benefit aspect and social cum environmental benefits per annum suggest that return on investments for social and technological capital invested has a positive impact and indicates the allied benefits as discussed earlier. The success of the innovative approach also depends on the reliability and recreation of the innovation.

Lessons learnt

The study was an attempt to understand key dimensions of sustainability at a business school campus a particularly from an operations point of view. This was important as it helped justify the existing cultural ethos and approach towards the journey of sustainability and without relevant policies, other sustainability measures would remain only superficial and not efficient even with change of staff (Chambers, 2009).

However the study also revealed a few interesting aspects of the approach taken on the path to sustainability. This would be helpful for other campuses and Universities to consider as part of regulating campus operations including compliance with national education regulators for efficient functioning of Universities.

One of the major lessons learnt from this study was to consider the inclusivity and integration of several departmental units and readiness for process standardization. This also meant capacity building of human resources to handle the progression to an online system and subsequent monitoring.The prerequisites to implement such an e process system at an educational institute should consider the following aspects:

1. Stakeholder satisfaction

High student satisfaction in effective, reliable and timely operation processes was seen as critical to the overall academic integrity of the institute since students were considered as major stakeholders. Although this study did not analyse the satisfaction levels of faculty and administration staff for transition to an online process system, it was apparent that an efficient online academic process was widely accepted as a sustainable approach by all.

2. Infrastructure

The availability of reliable, efficient and user friendly technology in terms of hardware and related software should be available to implement the ICT based systems and processes.The infrastructure should also include sufficient band width for reaching out to the large student communities staff and administration as well as access to information systems and easy availability of the applications.

3. Skilled Workforce

It is essential to have a skilled workforce to manage the ICT enabled processes. The training of IT personnel and implementation of an ICT based approach plays a crucial role in implementation of the process including monitoring and control of systems particularly during the academic sessions and admission process.

4. Support from Top Management

Any change management initiative in ICT with coherent policy and support from higher and senior level management (in terms of capital investments and workforce investments) will further add value to operationalising the transition to an online system.

5. Process standardisation

It is often suggested that change is the only constant and perhaps also the most difficult to accept for an existing state of established group or individual. In such a scenario changing mindsets, behaviour patterns and readiness for process standardization of an innovative on-line system can be a major barrier to implementing a process. The transition must therefore take into account the development of a strategic framework (Hart and Milstein, 2003) and action plan (Calder and Clugston, 2003) which will be able to develop an effective and robust 7 by 4 University Sustainability Framework.

Conclusion

The present study was one of the first initiatives to understand the effectiveness of developing an online process based system in a traditional business school environment. The transition to an electronic and online based academic process was for the first time attempted in the model at the Symbiosis Institute of International Business, Pune where hitherto manual based sys-tem were being implemented ever since the Institute was founded in 1992.

Taking into consideration the criticality of the administrative process and also with the in-troduction of Right to information (RTI) act in India, the present efforts in developing an online based administrative system to monitor academic processes will be of great benefit for improved efficiency. The earlier manual process was mainly person dependent and time consuming. The ICT based process improvements is a sustainable solution and could be replicated at other educational institution also.With the present online system the administra-tive process has evolved and progressed from a person dependent to process dependent ap-proach with the help of ICT. The process model has saved not only time and become cost effective but also led to development of sustainability as a part of the organisation culture of the institute.

References

Barth M., Godemann J., Rieckman M., Stoltenberg U (2007) Developingkey competences for sustainable development in highereducation. *International Journal of Sustainability in Higher Education* 8 (4), 416-430.

Blackburn, W. (2009), Handbook of Sustainability, Pages-343-343, Greenleaf publishing.

Bandalaria, M. (2007), Impact of ICTs on Open and Distance Learning in a Developing Country Setting: The Philippine experience. *International Review of Research in Open and Distance Learning,* 8 (1), 1-15.

Carbon Trust (2006), "Higher Education Carbon Management Programme", *http://www.car bontrust.co.uk/carbon/he/* (November 2006).

Calder, W. and Clugston, R.M. (2003) Progress toward sustainability in higher education. *Environmnetal Law Reporter* Vol 33 No.1, pp. 10003-10023.

CasulaVifell, Å. andSoneryd, L. (2012) Organizing matters: how the "social dimension" gets lost in sustainability projects. *Sustainable Development* 20 (1), p. 18-27.

Chambers, D. (2009) Assessing and planning for Environmental sustainability – A framework for Insitutions of Higer education In Leal Filho, W. Eds (2009), *Sustainability at Universities: Opportunities, Challenges and Trends, Series: Umweltbildung, Umweltkommunikation und Nachhaltigkeit / Environmental Education, Communication and Sustainability*, Vol. 31, Frankfurt, Peter Lang Scientific Publishers, p. 287-297.

Copernicus guidelines for sustainable development in the European Higher Education Area (2010), p. 41.

Cloquell-Ballester V.-A., Monterde-Diaz R., Cloquell-Ballester V.-A., Torres-Sibille A.d.C., 2008, "Environmental education for small – and medium-sized enterprises: methodology and elearning experience in the Valencian region" *Journal of Environmental Management* 87 (3), p. 507-520

De Kraker, J., Lansu A., Dam-Mieras R. van (2007): Competences and competence-based learning for sustainable development. In: De Kraker, J, Lansu, A, Dam-Mieras R. van: *Crossing Boundaries. Innovative Learning for Sustainable Development in Higher Education*. VAS: Frankfurt a.M., p. 103-114

Economist Intelligence Unit, (2008), The future of higher education: How technology will shape learning, p. 29.

Eneroth, C. (2000), E-Learning for Environment. Improving e-Learning as a Tool for Cleaner Production Education. 236 p. Licentiate Dissertation. Lund University, Sweden

Garrison, R. (2000),Theoretical Challenges for Distance Education in the 21st Century: A shift from structural to transactional issues. *International Review of Research in Open and Distance Learning*, 1, p. 1-17.

Godemann, J., Herzig, C., Moon, J. and Powell, A. (2011), "Integrating Sustainability into Business Schools – Analysis of 100 UN PRME Sharing Information on Progress (SIP)" reports, No. 58-2011 ICCSR Research Paper Series, (Editor: Jeremy Moon), Research Paper Series, International Centre for Corporate Social Responsibility.

Hart, S.L. and Milstein, M.B. (2003) Creating sustainable value. *Academy of Management Executive*, 17(2), p. 57-67.

Higher Education in India: Twelfth five year plan (2013-2017) and beyond, FICCI report. Assessed VIA *www.ficci.com* on 25th October 2013.

Indian Higher Education Sector Opportunities aplenty, growth unlimited, Deloitte Publishing, October 2012.

Leal Filho, W. Eds. (2009), Sustainability at Universities: Opportunities, Challenges and Trends, Series: Umweltbildung, Umweltkommunikation und Nachhaltigkeit / Environmental Education, Communication and Sustainability, Vol. 31, p. 338, Frankfurt, Peter Lang Scientific Publishers.

Leal Filho, W. (2011), "About the Role of Universities and their Contribution to Sustainable Development" *Higher Education Policy*, 24, p.427-438.

Leal Filho, W. (2012), (Ed), Sustainable Development at Universities: New Horizons Series: Umweltbildung, Umweltkommunikation und Nachhaltigkeit / Environmental Education, Communication and Sustainability, Volume 34, p. 994. Frankfurt, Peter Lang Scientific Publishers

Marcuse, P. (1998) Sustainability is not enough. *Environment and Urbanization* 10 (2), p. 103-111.

Mihane, B. and Badivuku-Pantina M. (2010), "The impact of new technologies in the knowledge society", *Lex ET Scientia International Journal – Juridical Series*,Vol (1), p. 449-455

Mooij, T. (2009), Education and ICT-based self-regulation in learning: Theory, design and implementation, *EducInfTechnol* 14, p. 3-27

Murphy, K. (2012) The social pillar of sustainable development: a framework for policy analysis. *Sustainability: Science, Practice & Policy* 8 (1). p. 15-29

National Skill Developmental Report (2012) assessed on 25th Oct 2013 via *www.nsdcl.com*

Naeem, M. and Neal, M. (2012), "Sustainability in business education in the Asia Pacific region: A snapshot of the situation". *International Journal of Sustainability in Higher Education*, 13(1), p. 60-71.

Pandey, D., Agrawal, M., and Pandey, J.S. (2011), "Carbon footprint: current methods of estimation", *Environmental Monitoring and Assessment,* 178(1-4), p. 135-160.

Petrides, L and Nodine,T R (2003), Knowledge Management in Education ,The institute for the study of knowledge management in education, p. 30.

Porter, M. (1990), The competitive advantages of Nations, Harvard Business Review, March-April, 1990.

Prahalad, C.K. (2006), The Fortune at the Bottom of the Pyramid: Eradicating Poverty Through Profits, Dorling Kindersley Pvt Ltd, p. 432.

Rao, P. (2011), "Integrating sustainability into global business practices – an emerging tool for management education", In *Internationalisation of Higher Education* (Eds). Rajani-Gupte, B Venkataramani, and D Gupta), p. 60-73. Excel Books.Publ.

Roy, R., Potter, S. and Yarrow, K. (2008), "Designing low carbon higher education systems: Environmental impacts of campus and distance learning systems", *International Journal of Sustainability in Higher Education,* 9(2), p. 116-130.

Srivastava, S. (2010), Technology Can Change the Education Landscape. *http://www.edu-leaders.com/content/technology-can-change-education-landscape* accessed on 29 October 2013.

Spranger, S. (2011), "Calculating the carbon footprint of Universities", Master's Thesis, Economics and Informatics, Erasmus School of Economics, Unpublished Thesis, p. 107.

Tamm Hallström, K. & Boström, M. (2010) Transnational Multi-stakeholder Standardization: Organizing Fragile Non-state Authority. Northampton, MA: Edward Elgar.

University of Toronto (2010), St. George campus greenhouse gas emissions inventory report.

IV.
Formal and non-formal case studies/experiences

Building an Online Master's Program for Deep Learning in Sustainability

Amelia Clarke[1]

Abstract

Deep learning is seen as a way to maximize the benefits from sustainability education for both students and society. This chapter considers the University of Waterloo's Master of Environment and Business (MEB) program in relation to seven characteristics of deep learning. The MEB program is an executive education program that is mostly delivered online. Student survey responses show a high satisfaction with the program and highlight areas where deep learning is occurring. This chapter emphasizes that it possible to ensure deep learning in an online program and in a program's design (not just at the teaching activity/ assignment level). The chapter ends by offering lessons learned for other online courses and programs.

Introduction

Sustainability education is a seen as a way to educate students about environmental and social concerns, while at the same time supporting a transition towards sustainability in society (McNamara, 2010). Deep learning is particularly relevant for education for sustainability due to the interdisciplinary nature and holistic insights needed (Warburton, 2003) and the desire to actually change organizations and systems (Fullan, 2006; Stirling, 2004). Deep learning is usually defined in distinction to surface learning; a deep approach to learning as one in which students "seek meaning in order to understand it" (Trigwell and Prosser, 1991, p. 251).

The Master of Environment and Business (MEB) program at the University of Waterloo is designed for working professionals. Following a two-week residency period, the rest of the degree is online. Students come from different academic backgrounds, different workplaces, and different countries. What they have in common is a desire to build more sustainability related content into their existing or future job. Given that the program was designed from scratch, there was a great opportunity to prepare the curriculum and delivery to ensure deep learning.

This chapter details how the MEB program at the University of Waterloo encompasses seven characteristics of deep learning. The literature review explains deep learning. As context, the 'environment and business' degree, program design, course attributes, and student characteristics are explained. A table demonstrating how the seven key characteristics of deep learning are built into the MEB program is also presented. Throughout the results section, quotations which have been obtained from anonymous student evaluations demonstrate the value of specific assignments to students. Finally, lessons learned for other online courses and programs interested in deep learning are offered.

---

1   University of Waterloo, Waterloo, ON, Canada, N2L 4V5.

Literature Review on Deep Learning

Deep learning is short for deep approaches to learning, and differs from a surface learning approach. A surface approach to learning is defined as one approach where students memorize mechanically in order to reproduce learning materials (Trigwell and Prosser, 1991). By contrast, deep learning enables students seek deep meaning and understand learning materials (Trigwell and Prosser, 1991). A Task Force on Innovative Teaching Practices to Promote Deep Learning at University of Waterloo identified seven characteristics which exemplify students' application of deep learning (Ellis et al., 2011):

1. retain knowledge and apply it to new and different contexts (also known as transfer);
2. relate ideas and make connections between new and prior knowledge;
3. see concepts, ideas and/or the world differently;
4. engage in independent, critical, analytical thinking in a quest for personal meaning;
5. regulate themselves as learners;
6. rely on intrinsic motivation to learn; and
7. engage in active learning by interacting with others and the course material.

Research suggests that when students take a deep approach to learning, they are able to better transfer knowledge (Ellis et al., 2011). Related to the first two characteristics of deep learning, Halpern and Hakel (2003) identify ten basic principles that effective teachers can use to retain and transfer knowledge: 1) ensuring practice at retrieval; 2) varying the conditions under which learning takes place; 3) requiring learners to take information and present it in one format and re-represent it in an alternative format; 4) building on prior knowledge and experience; 5) recognizing students' and instructors' own epistemologies; 6) recognizing that experience alone is a poor teacher; 7) remembering that lectures work well for learning assessed with recognition tests, but work badly for understanding; 8) noting that the act of remembering itself influences what learners will and will not remember in the future; 9) being aware that less is more, especially when we think about long-term retention and transfer; and 10) embracing that what learners do determines what and how much is learned, how well it will be remembered, and the conditions under which it will be recalled (Halpern and Hakel, 2003).

The third characteristic of deep learning is students "come to see concepts and the world differently" (Ellis et al., 2011, p. 7). Stirling (2004) equates the level of learning with the level of systemic change possible. Single-loop learning (or surface learning) does not change the values or perspective of the student. Double-loop learning, in contrast, is characterized by a change in frame which also leads to second-order change, which is a change in the system (Stirling, 2004). Transformative learning, which Stirling (2004) equates to a third level even deeper than double-loop learning, is based in systems thinking and can actually lead to a paradigm shift. The Ellis et al. (2011) framework considers anything deeper than single-loop learning to be deep learning.

The fourth characteristic – "engage in independent, critical, analytical thinking in a quest for personal meaning" (Ellis et al., 2011, p. 7) – is typical of any graduate level program. For example, the Graduate Degree Level Expectations in Ontario Canada (University of Waterloo, 2013b) concentrate on "a systematic understanding of knowledge and a critical awareness of current problems and new insights, a conceptual understanding and methodological competence, competence in the research process by applying an existing body of knowledge in the critical analysis of a new question or of a specific problem or issue in a new setting, [and] the ability to communicate ideas, issues and conclusions clearly" (University of Waterloo, 2013b, p. 1).

The fifth characteristic of deep learning is that students "regulate themselves as learners" (Ellis et al., 2011, p. 7). Self-regulation is defined as "not a mental ability or an academic performance skill; rather it is the self-directive process by which learners transform their mental abilities into academic skills" (Zimmerman, 2002, p. 65). Self-regulation realizes the life-long goal of education as developing learning skills (Zimmerman, 2002). Research on self-regulation implies that students who are seldom given choice in study have difficulty in developing self-regulated learning skills (Zimmerman, 2002).

The last two characteristics of deep learning are: students "rely on intrinsic motivation to learn" and "engage in active learning by interacting with others and the course material" (Ellis et al., 2011, p. 7). Intrinsic motivation is defined as "the motivation for ongoing interaction with the environment" (Deci et al., 1981, p. 1). Additionally, students' ability and motivation have deep impacts on the quality of learning interactions (Kawachi, 2003). By being motivated to understand and engage in the topic, they will pursue deep learning (Warburton, 2003). Waterburton (2003) offers three factors that influence this motivation: the learning environment (such as the online platform, teaching style, and opportunity for discovery); the course content (such as key concepts and themes, range of content, and personal relevance); and individual factors (such as prior experience, workload, and background knowledge).

There is a literature that considers deep learning that is highly relevant for online programs. In particular, teaching methods are emphasized. For example, a high level of communication and interactions among professionals and students contribute to effective teaching (Offir et al., 2008). Also, students' reflections have significant potential to promote deep learning (Lynch et al., 2012). Self and peer evaluations can lead to better critical thinking and learning outcomes (Lynch et al., 2012).

On a whole, integrated online course models can promote learning satisfaction and reinforce knowledge (Ke and Xie, 2009). Online discussion can help teachers deepen teaching methods (Lee and Baek, 2012), promote deep learning and have positive impact on memory rates (DeLotell et al., 2010). Similarly, a framework for deep learning in online discussion indicates that dynamic online interaction promotes students' learning skills and leads to deep learning (Du et al., 2005).

Specific to sustainability, the value of deep learning is well understood given the complexity of the topics and the desire to support the transition to a sustainable society (Warburton, 2003; Stirling, 2004; Fullan, 2006). Effective education for sustainability will ensure student reflection, discovery and action research (Warburton, 2003). Involving students in professional learning communities has also been encouraged (Fullan, 2006). Stirling (2004) offers three levels of education & sustainability: 1. education about sustainability which leads to bolt-on changes; 2. education for sustainability which leads to building in change; and 3. sustainable education which leads to rebuilding and redesigning.

While the literature states that reflections, self and peer evaluations, interacting with professionals, class discussions, and integrated modules are important, learning in the MEB program design has many more features, such as content delivery methods, assignment formats, team presentations, group reports, individual assignments related to workplaces, etc. There has also been considerable thought in the MEB design to ensure the students are able to enact change. This chapter, focused on deep learning, highlights how considering deep learning at the program design level instead of the individual activity or assignment level, ensures that the degree itself enabled a deep approach to learning.

Context: Master of Environment and Business (MEB) Program

The Master of Environment and Business (MEB) is an online program "aimed at meeting the growing need for business sustainability professionals as a distinct group of knowledgeable, skilled, confident and motivated individuals with the information, tools and expertise to integrate environment with business in very practical ways" (University of Waterloo, 2013a, p. 1). Graduates are prepared for senior level and strategic positions in businesses, or in other organizations seeking to change (or influencing others to change) to more sustainable practices. "The MEB program offers students a MBA-equivalent degree through online courses with minimal on-campus study, and with course materials distributed over the Internet" (University of Waterloo, 2013a, p. 1).

As the students are working professionals, most take the degree as a part-time program of study, with one course a semester. At this pace, it takes three years to complete the ten courses and two milestone conferences. In total, there are approximately 85 students in the MEB program. The first cohort of MEB students graduated from the University of Waterloo in October 2013. For more information about the program design see Table 1 and also the program website[2].

Table 1: MEB Program Design

	Course Topics	Purpose
Foundation Courses **(2 courses)**	▪ Business Case for Sustainability ▪ Sustainability for Business	▪ Know each other, professors and program ▪ Common knowledge base ▪ Academic skills ▪ Teamwork and online skills
Core Courses **(4 courses)**	▪ Green Marketing and Social Accountability ▪ Business Operations and Sustainability ▪ Strategies for Sustainable Enterprises ▪ Environmental Finance	▪ Deeper knowledge ▪ Introduction of content-based skills
Electives **(2 courses)**	▪ Enterprise Carbon Management ▪ Lifecycle Assessment and Management ▪ Stakeholder Engagement, Collaboration and Partnerships ▪ Sustainability Reporting	▪ How-to courses ▪ Deeper content-based skills
Capstone **(double course)**	▪ Capstone Sustainability Project	▪ Applied project-based course; including a major research paper
Milestone Conferences **(must attend two)**	▪ Professional conferences with additional MEB side events	▪ Learn leading edge conversation ▪ Interact with professionals

---

2   For more information see: uwaterloo/env/MEB.

Results

The MEB program was designed from scratch, and thus considerable thought was put into the design. The student and topic attributes lend themselves to deep learning in many ways. As the students are working professionals, this enables the use of authentic evaluation strategies and real world contexts. Assignments can be applied to their workplace, and the diversity of students ensures different perspectives are represented in class discussions. For example, some students work in large corporations, others in smaller companies, some in the non-profit organizations and others in the public sector. All have at least three years of work experience prior to starting in the program. There is also a wide variety of industries represented. Here is what one student had to say in an anonymous survey about his/her favorite part of the program:

> "... meeting a diverse group of peers and professors, which leads to interesting discussions and varied points of view"

The 'environment and business' topics also provide opportunities to bring practical real-world content into the classroom. For example, the online platform allows the instructor to link to live websites for current case studies made up of video, web content, and even guest expert question and answers (on a discussion board). The students are challenged to apply critical thinking to the 'real-world' content they are reading. This quotation from the anonymous survey from a student about his/her favorite part of the program highlights the real-world content in the program:

> "I do find the content relevant and often tied to case studies which I enjoy. There is no doubt that the professors are leaders in this field and there are lots of real world evidence and exploration offered to students."

The MEB course and milestone conference attributes also have deep learning embedded in their design. For example, there is considerable choice in assignments and opportunity for students to self-direct their weekly engagement as it is asynchronous. Each course has an individual assignment and a team assignment; some of these allow for students to tie them to their workplace. For example, in the Introduction to Sustainability for Business course, students write a hypothetical memo to their CEO (or senior person). Some students are actually able to make use of this content for other work purposes. The final capstone project is a research project with a client, but which also must make an academic contribution. This quotation from the anonymous survey question about his/her favorite part of the program highlights how applicable this student is finding the content:

> "... the knowledge I gained regarding sustainability and what it means for organizations, from the fundamentals to the more specific details. A little more than half way through the program, I feel confident in my abilities to apply the knowledge I have gained in this program and have a successful career as a business sustainability professional."

Another design feature of every course is that they all have interaction with professionals about the course material. Sometimes this is done by pre-recording content, or through discussion board question & answer sessions. One of the advantages of the online format is the guest can be from anywhere in the world. Also, all of the MEB instructors have some practical experience that they are able to bring into the classroom. Every course also has student-led discussion boards and other activities. The milestone conferences require the students to participate in an MEB orientation, a social event, the main conference, and also to write a reflection. One student mentioned in the anonymous survey this comment about the guest expert discussion boards:

"The expert discussions were an unexpected added bonus. It was such a privilege to interact with some of these experts and get the inside story on their work. [The guest experts] made the case studies so much more relevant and addressed the nuances not apparent in reading the websites."

In terms of the discussion boards, in each course these are tied to the weekly or biweekly content. A new board begins every week or two. This paces the students during the course, but also enables the conversation to stay active and for different students to lead different boards. This was an unsolicited comment that was posted on a discussion board in a course by a student to his/her peers:

"Thank you dear classmates for all your contributions. I thought it was time – during November, the darkest month and half way through our three year program – to send a note of thanks for sharing your experience, knowledge and insights. I am gaining so much from reading your views on our course content. I so appreciate reading your posts and wonder how we can all look at the same topic so differently. I've come to count on [these discussions for] expanding my knowledge."

In terms of the online nature of the program, students tended not to comment on this as there was no in-person option. In general, students chose this online program so that they could work while in school and so that they could be located anywhere. Here is what one student had to say to a question about his/her favorite part of the program:

Of most benefit has been the format – I would not be able to obtain this degree otherwise. The online format allows me to study abroad and interact at the level necessary to feel engaged in a valuable learning process.

Using the deductive framework provided in Ellis et al. (2011) about deep learning characteristics, an analysis was conducted about the Master of Environment and Business design. Table 2 highlights these results.

Table 2: Key Characteristics of Deep Learning in MEB Program.

Deep Learning Characteristic (Ellis, D., et al., 2011)	MEB Design Features
Retain knowledge and apply it to new and different contexts	• Applied assignments and capstone project • Assignment formats, for example different teams focusing on different industries
Relate ideas and make connections between new and prior knowledge	• Program design builds on previous courses and experiences
See concepts, ideas and/or the world differently	• Course content • Student diversity • Negotiation exercise
Engage in independent, critical, analytical thinking in a quest for personal meaning	• Individual assignments • Reflections
Regulate self as learner	• Asynchronous design • Choice in assignments • Additional resources in each week's content
Rely on intrinsic motivation to learn	• Tied to workplace • Milestone conferences
Engage in active learning by interacting with others and the course material	• Discussions • Team presentations and reports • Engage professionals

In summary, the MEB program is an online program which aims to facilitate deep learning on business and sustainability. In the MEB program, some assignments are designed to apply concepts in different contexts, and to build off prior knowledge. Also, student diversity helps students see concepts and ideas from different angles. Individual assignments, reflections and discussions increase students' engagement in independent, critical, and analytical thinking. Online assignments in the MEB and options available regarding assignments also enable students to regulate themselves as learners. The entire program only functions if students are intrinsically motivated to learn; so this is a criteria used as part of admission. In addition, the MEB has numerous interaction opportunities.

Discussion and Conclusion

While these quotations were extracted from a survey to provide examples regarding deep learning, they do reflect the positive attitude that students have towards the program. In the 2013 survey – which had a 58% response rate – 92% of the students were satisfied with the MEB program and 100% were likely to recommend it to others. The main area for frustration for students about the program design is in regards to the virtual teamwork in every course. It is not surprising that some students do not like group assignments, but the virtual element (with students in different time zones) on top of working full-time makes these assignments harder to manage time-wise.

Using Waterburton's (2003) three factors that influence deep learning as a means of organizing this section, the following are some lessons learned through designing and implementing the MEB program.

The learning environment

Designing a program from scratch provides numerous opportunities to ensure deep learning at the program level. Much of the deep learning opportunities come through the learning environment. The online course designs, including the assignments, are a large part of that where deep learning can be structured into the program. For the MEB this includes the applied assignments, exercises, reflections, discussions, teamwork and guest experts. The students love the assignments that are applied to their workplaces and/or where they have significant choice. That said, for these assignments, effective feedback on these assignments from the instructor is critical for the student's learning. It requires the professors to be very comfortable with the material and the field to be able to comment on such varied submissions and provide meaningful feedback on the content.

Online discussions are an important tool for online courses to promote deep learning (Lee and Baek, 2012; DeLotell et al., 2010; Du et al., 2005). In terms of the discussion boards, from both a student and instructor perspective, these are hard to 'time budget'. The more motivated the students are, the more posts they make, and the more they suggested additional links for their classmates to explore. Students who come into the conversation later in the discussion (due to work or home obligations), but before the deadline, can have trouble engaging in the conversation. Also, for the guest experts who log into the discussion board for the question and answer session, the number of questions waiting can be overwhelming. Even with careful discussion assignment restrictions, ensuring deep learning by all students through this activity can be challenging.

Interaction with professionals by the students is important (Offir et al., 2008, Fullan, 2006). In the MEB program, this is not only done through the classroom and applied assignments, but also through milestone conferences. This mandatory degree requirement has added considerable value to this professional Master's degree program. The students learn the leading edge conversation, network with a large number of professionals, interact with each other (across cohorts), and see how the movement networks.

Course content

The MEB was designed with both skills and curriculum in mind. The attributes of the graduating students and the knowledge to be learned were determined. Then the content to be covered at an introductory, medium and high level were mapped across the courses, ensuring that the foundation courses covered the base knowledge and skills needed by all students (thus accommodating various experience and training backgrounds), while the core courses went deeper and the electives and capstone ensured a 'how to' level, including application. Compared to Stirling's (2004) levels of education & sustainability, the early courses teach the education about sustainability, while the rest teach the education for sustainability and the sustainability education. Students are challenged not only to know bolt-on solutions, but also how to build change from within, and redesign systems. They also learn what has worked (and not worked) and the challenges of being the change agent. For those who can directly tie the capstone project and other course assignments to their workplace, sometimes they apply course learnings to creating change immediately. The beauty of executive education is the opportunity for the students to already be in positions to effect change directly.

Now, as the MEB program starts to build the alumni network, and decide how best to leverage cross-cohort connections for larger transformational change, there are limitations. It is not part of the mandate of the MEB Director; rather the role is focused on recruitment/admissions, program improvements/maintenance and current student development. Alumni services are elsewhere in the institution, and more focused on donor relations. Building the alumni network and services to reach its potential is beyond the current capacity of the program team. That said, the MEB team is currently experimenting with easy-to-implement options.

Another note about course content is that business and sustainability is a dynamic field that keeps evolving. It is important to keep the courses current, even if most of the material is pre-prepared. The online environment has created some challenges in regards to staying current as all courses are prepared months in advance of offering. These challenges can be overcome through announcements and discussion posts.

Individual factors

Perhaps of most importance to the MEB program design is the selection of students. These professional students gain about 50% of their learning from each other. Having students with at least three years work experience, and a direct interest in matching sustainability to their career have been important admission criteria that have led to the success of the program. The average age is 35, with a mix of junior, middle and senior managers. For example in each cohort, there have been some executive directors, chief executive officers and/or managing partners. About half of the class already works in sustainability-oriented jobs, including as sustainability coordinators, senior policy advisors, environmental officers, etc. The other students have relevant experience to offer the discussion on the management side, even if they are newer to the sustainability topics. These individual attributes enable deep learning.

In conclusion, the results provided in this chapter emphasize that it possible to ensure deep learning in an online program. They also provide evidence that it is possible to consider deep learning at the program level and not just at the teaching method/activity/assignment level.

Aknowledgements

The author would like to thank Wen Tian for her RA work on this topic.

References

Deci, E.L., Nezlek, J., and Sheinman, L. (1981), "Characteristics of the Rewarder and Intrinsic Motivation of the Rewardee", *Journal of Personality and Social Psychology,* 40(1), 1-10.

DeLotell, P.J., Millam, L.A., and Reinhardt, M. (2010), "The Use of Deep Learning Strategies in Online Business Courses to Impact Student Retention", *American Journal of Business Education,* 3(12), 49-55.

Du, J., Havard, B., and Li, H. (2005), "Dynamic Online Discussion: Task-Oriented Interaction for Deep Learning", *Educational Media International,* 42(3), 207-18.

Ellis, D., Bissonnette, C., Furino, S., Hall, S., Kenyon, T., McCarville, R., Stubley, G., and Woudsma, C. (2011), "The Task Force on Innovative Teaching Practices to Promote Deep Learning at the University of Waterloo: Final Report." Waterloo, ON: University of Waterloo.

Fullan, M. (2006), "The Future of Educational Change: System Thinkers in Action", *Journal of Educational Change,* 7, 113-22.

Halpern, D.F., and Hakel, M.D. (2003), "Applying the Science of Learning to the University and Beyond: Teaching for Long-Term Retention and Transfer ", *Change,* 35(4), 36-41.

Kawachi, P. (2003), "Initiating Intrinsic Motivation in Online Education: Review of the Current State of the Art.", *Interactive Learning Environments,* 11(1), 59-81.

Ke, F., and Xie, K. (2009), "Toward Deep Learning for Adult Students in Online Courses", *Internet and Higher Education,* 12(3), 10-10.

Lee, H.-J., and Baek, E.-O. (2012), "Facilitating Deep Learning in a Learning Community", *International Journal of Technology and Human Interaction,* 8(1), 1-13.

Lynch, R., McNamara, P.M., and Seery, N. (2012), "Promoting Deep Learning in a Teacher Education Programme through Self- and Peer-Assessment and Feedback", *European Journal of Teacher Education,* 35(2), 179-97.

McNamara, K.H. (2010), "Fostering Sustainability in Higher Education: A Mixed-Methods Study of Transformative Leadership and Change Strategies", *Environmental Practice,* 12(1), 48-58.

Offir, B., Lev, Y., and Bezalel, R. (2008), "Surface and Deep Learning Processes in Distance Education: Synchronous Versus Asynchronous Systems", *Computers & Education,* 51(3), 1172-83.

Stirling, S. (2004), "Higher Education, Sustainability, and the Role of Systemic Learning". In *Higher Education and the Challenge of Sustainability: Problematics, Promise, and Practice,* edited by Peter Blaze Corcoran and Arjen E.J. Wals, 49-70. Netherlands: Springer.

Trigwell, K., and Prosser, M. (1991), "Improving the Quality of Student Learning: The Influence of Learning Context and Student Approaches to Learning on Learning Outcomes", *Higher Education,* 22(3), 251-66.

University of Waterloo. (2013a), "Master of Environment and Business", University of Waterloo. Retrieved from *https://uwaterloo.ca/school-environment-enterprise-develop ment/academic-programs/master-environment-and-business.*

– (2013b), "O.C.A.V.'S G.D.L.E.S (Graduate Degree Level Expectations)", University of Waterloo. Retrieved from *https://uwaterloo.ca/academic-reviews/graduate-programs/ graduate-degree-level-expectations.*

Warburton, K. (2003), "Deep Learning and Education for Sustainability", *International Journal of Sustainability in Higher Education,* 4(1), 44-56.

Zimmerman, B.J. (2002), "Becoming a Self-Regulated Learner: An Overview", *Theory into Practice,* 41(2), 64-72.

First Online Course on Desalination by Renewable Energies, Lessons Learnt

Juan A. de la Fuente, Vicente J. Subiela and Baltasar Peñate[1]

Abstract

Potable water availability is one of the most pressing issues in the world and neither the over-exploitation of aquifers nor the uses of fossil fuels for desalination are sustainable solutions. In this chapter we describe a training initiative that was developed by the ITC under the framework of the PRODES project which was co-financed by the European Commission within the *Intelligent Energy for Europe Program*. The e-learning course called "Introduction to desalination by renewable energies" is addressed to those people who could be interested in this field of knowledge, as professionals related with water or energy sectors, or technology students. This e-learning course on desalination by renewable energies sets the fundamentals on renewable energy desalination as a contribution to address the world's water situation by increasing the knowledge on these technologies. After 12 editions made (5 within the *ProDes* project), the international impact is very relevant: 489 students from 35 different countries of the five continents. This course is offered in English, Spanish, French and Portuguese languages. After this experience, the ITC is committed to improve the course in each edition; an extended and upgraded version of the course (11th edition) has already concluded with the collaboration of the European Desalination Society (EDS).

Introduction

Sustainable development is a complex concept which concerns a wide range of social, techno-economic and environmental issues. Without addressing all these dimensions, teaching of sustainable development would not be complete. Therefore, taught modules and teaching materials for engineering students should include not only technological analysis and economic evaluation, but also environmental and social considerations.

The idea of Education for Sustainable Development (ESD) germinated through the report of the World Commission on Environment and Development (1987) entitled Our Common Future. The United Nations' Decade for Education for Sustainable Development or DESD (2005-2014) encompassed action themes, including overcoming poverty, achieving gender equality, health promotion, environmental protection, rural development, cultural diversity, peace and human security, and sustainable urbanization (UNESCO 2005). Also The World Conference on Education for Sustainable Development (2009) defined ESD as "*an approach to teaching and learning based on the ideals and principles that underlie sustainability*" including with key issues as "*human rights, poverty reduction, sustainable livelihoods, climate change, gender equality, corporate social responsibility, protection of indigenous cultures in*

---

1   Water Department, Research and Technological Development Division, Canary Islands Institute of Technology (ITC), Playa de Pozo Izquierdo, s/n. 35119 Santa Lucía, Gran Canaria, Spain
    Tel. +34 (928) 72 75 20; Fax +34 (928) 72 75 90.

an integral way, it constitutes a comprehensive approach to quality education and learning"
(retrieved from *http://www.esd-world-conference-2009.org/en/about-world-conference-on-esd/objectives.html*), and e-learning should provide "*a model for sustainability education among higher educational and governmental institutions covering comprehensive framework of sustainability, such as concepts, sustainable resource use institution, economy, social capital and human capital*" (from *http://portal.unesco.org/* New e-learning course: Sustainability Science)

In this chapter we describe a training initiative that was developed by the ITC under the framework of the PRODES project which was co-financed by the European Commission within the "*Intelligent Energy for Europe Program*". The e-learning course called "Introduction to desalination by renewable energies" is addressed to those people who could be interested in this field of knowledge, as professionals related with water or energy sectors, or technology students. This e-learning course on desalination by renewable energies sets the fundamentals on renewable energy desalination as a contribution to address the world's water situation by increasing the knowledge on these technologies. Evaluation of the first 12 editions is given and discussed.

The role of water & energy in sustainable development

The World Health Organization (WHO) has estimated that 1000 cubic meters per person per year is the benchmark level below which chronic water scarcity is considered to impede development and harm human health. Desalination systems are of paramount importance in the process of augmenting fresh water resources and happen to be the main life support systems in many arid regions of the world.

Desalination as currently practiced is driven almost entirely by the combustion of fossil fuels. It is an energy intensive process. These fuels are in finite supply and pollute the environment and interfere with the biosphere. With the heavy water stress that is expected to increase over the next few decades, more oil has to be spared merely to obtain water. There seems to be a marked lack of attention to these combined water and energy issues, which are of paramount importance in the context desalination. The orientation of desalination practice at present is directed away from sustainability.

For this reason desalination must be analyzed using the three elements of sustainability: cost, society, and the environment. Students must approach the desalination processes with sustainability in mind; understanding water production via desalination within the water-energy-cost nexus and knowing renewable-energy-powered desalination processes.

Promotion of Renewable Energy for Water production through Desalination

As mentioned before, fresh water supply in the world, particularly in developing countries, is becoming a more and more challenging problem and affects to multidisciplinary aspects, as security, health, development, economics and environment. The increment of population, the climate change and the environmental impacts on the water resources are generating a progressive reduction in the per capita drinking water availability, particularly in developing countries of Africa and Middle East. According to the WHO, child mortality related to lack of water quality or quantity is more than 1.4 million per year (source: WHO, Water Sanitation and Health, 2010).

The problem of water supply in dry areas boosted the development of desalination technologies, especially in the Middle East, wherein most of desalinated water is produced. Furthermore it has become a very significant industrial activity with a progressive growth; desalination technologies have been producing fresh water supply for more than five decades; the current world capacity desalinated water technology installed is over 73 million of daily cubic meters (installed membrane capacity: 49.9 million m^3/d. Installed thermal capacity: 23.2 million m^3/d. Source: GWI DesalData / IDA 2012). Nevertheless, this solution has an important disadvantage: the high energy consumption, as heat for distillation processes, or electricity for membrane processes, like reverse osmosis or electro dialysis. Heat and electricity are mostly generated from fossil fuels and thus the main environmental consequence is the impact in terms of CO_2 emissions (as a rough reference, each cubic meter of desalted water from seawater RO plant means a generation of about 3 kg of CO_2).

Desalination increases energy demand, so it means a higher external dependence and the consequent economic expense in those countries with low energy resources.

Desalination by renewable energies solves both issues: on one hand it is a free-pollution system; on the other hand, it uses a local energy source as solar radiation or wind energy to power the desalination plant. So we can state that this inconvenience can be solved currently for small water demands by the use of renewable energies, as solar or wind energy.

The *ProDes* Project in the Canary Islands

The Canary Islands is a region pioneer in these facilities, is an obvious example where desalination plays a key role in the fresh water supply.

The Canary Islands Institute of Technology (ITC) Water Department, with a large experience (since 1996) in testing desalination units driven by renewable energies (figure 1) implemented the first online course (*www.desreslearning.com*) focused on this topic. This training initiative was developed by the ITC under the framework of the PRODES project (*www.prodes-project.org/*) which was co-financed by the European Commission within the "*Intelligent Energy for Europe Program*".

The *ProDes* project, standing for "*Promotion of Renewable Energy for Water production through Desalination*" started on 1st October 2008 and ran for 2 years. This project brought together 14 leading European organizations in order to support the use of renewable energy for powering desalination in remote or isolated areas where the grid cannot accommodate high penetration of intermittent sources. The focus was in Southern Europe where desalination is an increasingly important energy demand factor. The project involved the industry, academia, investors and institutional actors to deal with the non-technical barriers, like:

✓ The lack of coordinated research and industrial product development on the European level;
✓ The lack of formal training and specialized personnel;
✓ Local actors are not up-to-date with the technological developments;
✓ Funding for product development and project implementation is not easily available;
✓ The general public is not aware of the benefits renewable energy desalination has to offer.

Introducing RE-desalination to higher education and training actions

ProDes used a mix of tools like specialized courses, workshops, publications and others to deal with the issue. Within work package (WP 3) "*Introducing RE-desalination to higher education*", various courses were implemented in Greece, Spain, Italy and Portugal.

The main objectives of these training actions were:

➢ To introduce RE-desalination in the higher education system of relevant countries, in order to fill the knowledge gap and help produce the missing specialists that will work with entrepreneurs that are active in this fast emerging market;
➢ To implement courses for professionals, in order to deliver faster results by training the people that was already active in the market;
➢ To offer a basic training tool available for many people. That was the origin of the e-learning course, which could reach a much wider audience than the specific courses and according to experts on training, e-learning was becoming a more frequent educational option (Benchicou et al., 2010).

Figure 1: PV-RO by ITC in Tunisia.

The e-learning course "*Introduction to desalination by renewable energies*" was part of the training activities of the ProDes project and was implemented in the Moodle (**M**odular **Ob**ject **O**riented **D**ynamic **L**earning **E**nvironment) Platform (figure 2).

This e-learning action is based on an interactive and friendly use philosophy: the purpose is that on-line students assume the main role of their own training process in a flexible way. It means that students with high time restrictions are able to complete the course with a minimum dedication of 1 h/day, and on the other hand, students with more time availability can complete their training by the complementary options. Moreover, students can know their progress at any moment.

The course consists of ten chapters and the associated forums and quizzes, as the mandatory sections. Videos, games, podcasts and links to news, are complementary options.

Figure 2: General view of the course platform.

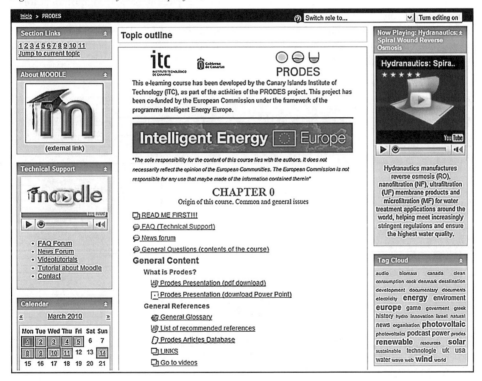

Course contents

The course is developed in ten chapters (9 theoretical + 1 practical case) having also several intermediate questionnaires for the evaluation. The theoretical chapters deal with the main relevant aspects of desalination (membrane and distillation processes) and the application of renewable energy technologies for their autonomous operation. Within the 9 theoretical chapters, there are 8 technical chapters and one specific chapter called "*Non-technical aspects*" focused on economic, environmental and social issues.

The main contents of the course are:

- Basic concepts on desalination and renewable energies;
- Desalination I. Membrane processes (EDR, RO);
- Desalination II. Distillation processes (MED, MSF, HD, MD);
- Solar thermal energy and MED;
- Solar thermal energy coupled to HD or MD;
- RO systems powered by PV solar energy;
- RO systems powered by wind energy;
- Other technologies;
- Non-technical aspects;
- Case study (practical case).

The practical case is an easy and friendly use tool to calculate the basic design of an autonomous desalination system (figure 3), allows the student to run a basic & easy use simulation tool to evaluate the main technical and economic parameters of four DES-RES combinations: photovoltaic driven reverse osmosis (PV-RO), wind powered RO, solar thermal multi-effect humidification (MEH), solar thermal membrane distillation (MD).

Figure 3: PV-RO practical case Excel data sheet.

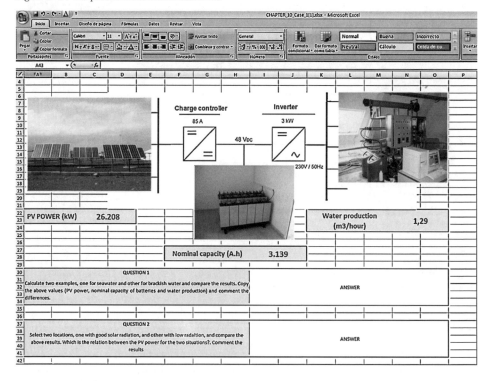

Lessons tool: each theoretical chapter is divided into 4 lessons and a review; a short question allows the student to pass to the next lesson.

The **Quiz** tool consists of multiple choice questions, one per theoretical chapter.

The **Forum** tool is the interactive part of the course; it is what enriches the course facilitating the experiences and knowledge exchange between the students, making the course more enjoyable and interesting and avoiding being simply an online book. The idea is to open a collective dialogue after ending each chapter to exchange points of view and specific knowledge; thus participation is a key factor.

The mandatory parts that all students must complete are: the 9 theoretical chapters, the 9 quizzes, one practical case and the participation in the forum of each chapter (at least one participation is required).

If students do not understand a technical term in the lesson, they can consult the **glossaries** for definitions of the main concepts of each chapter.

For **complementary training** students can go to the sections of **videos**, **podcasts**, **news**, visit the **suggested links** and **games**, based on the glossaries.

Learning methodology

A few days before the course starts, all the students receive by e-mail a document called "*user guide for students*" that will guide them through:

- Login on the platform;
- How to edit their profile;
- Page layout including blocks;
- Messaging;
- Forum;
- Activities and resources;
- Games.

CHAPTER 0 (common and general issues)

Students start the course reading the general issues in the section called "*READ ME FIRST*"; where they can find the first instructions and general issues related to this e-learning course.

In this introductory chapter students find, the section "*GENERAL FORUMS*" with three different forums:

One called "*FAQ*" used for the technical support with the Moodle platform use. The "*General Questions*" forum for all those doubts about the course contents. The "*News Forum*" is only for information about relevant events.

Once they familiar within the platform, they start the theoretical chapters in consecutive order; the platform opens one chapter per day. Firstly the student reads the first lesson of each chapter; a series of pages are presented sequentially, like a slide show (see figure 4). The lessons are scored with the use of intermediate questions for a grade.

After completing the lessons, the student must complete the questions of the quiz. Students will only have one opportunity to complete the quiz (a second chance will be given only for those who failed in their first attempt).

Each chapter is finished with the participation in the corresponding forum, wherein from a discussion topic /question the students will share their ideas, opinions and doubts also. This tool allows a mutual enrichment, very useful not only for the students but also for the teachers. On the other hand, students can open new discussion topics.

In summary, the proper order students should follow in order to complete each chapter is:

- Summary
- **Chapter**
- Glossary
- **Quiz**
- **Forum**

Students can click on the *Summary* to read an introduction about the Chapter, and use the *glossary* to know a little more about each topic (they are optional). The Chapter, questionnaire (quiz) and forum are **mandatory activities**.

The practical case (chapter 10) is the last mandatory element, which complements the theoretical training with a simulated situation in which the student evaluates the water supply by identifying the main design parameters of an autonomous desalination system in a remote location.

Figure 4: View of one slide taken from a lesson.

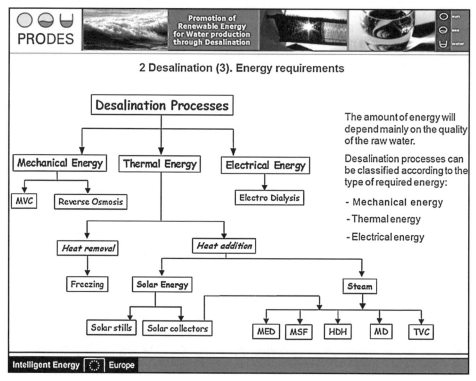

The course is completed with additional information as a list of recommended references, a publications database, suggested links, podcasts and videos.

Use of complementary elements

There are several complementary training options for those students with more time or specific interest. These optional activities include: podcasts, videos, games, glossaries, visit to links, and other elements in the platform.

In order to learn in a constructivism form, different games with words related to desalination have been created and students can play as long as they want because there is no score.

Role of teachers and evaluation

This e-learning course is a tutor supported course. A teacher evaluates part of the students training process, particularly the intervention in the chapter's forums, replying the student's comments and answering the questions made. The teachers also give a mark to the practical case and an additional mark (up to 1 extra point) according to the participation of the student in the optional activities. The platform calculates automatically the final mark for every student.

After completing the course, all those students who have achieved a grade higher than five receive an attendance certificate, with their name, final grade and date.

Course evaluation and lessons learnt

After completing the e-learning course and with the aim of improving it in further editions, students are required to fulfill a questionnaire in order to evaluate the course (figure 5). The evaluation questionnaire is totally voluntary and anonymous. Students have to evaluate the different items of the course from 1 (very poor) to 5 (excellent).

Figure 5: Course evaluation questionnaire

After answering all the questions, students just have to click over "*submit questionnaire*" and their answers get registered in the e-learning platform.

The student's participation in this questionnaire was closed to 33%. Up to 100 responses were obtained in the first six editions (the mean marks obtained in each question are shown in figure 6).

Figure 6: Course evaluation questionnaire results

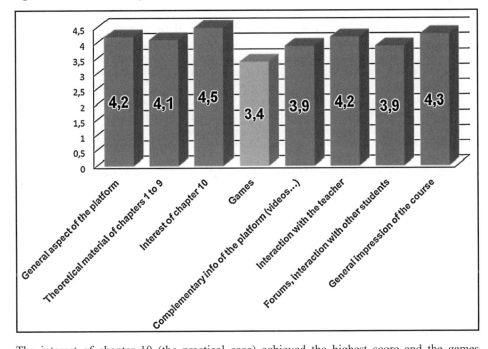

The interest of chapter 10 (the practical case) achieved the highest score and the games achieved the lowest score. The general impression of the course got 4,3 over 5.

Some of the observations made by the students in the evaluation questionnaire were:

- *Many thanks for the high level of **pedagogy** tools used in this course: the **clarity** by which the chapters, the quiz and the forums, were **structured** and the **good organization** of team managers (many thanks for their availability and sympathy) were the force points of this course.*
- *I think you offered a very **good introduction** to the basics of desalination, the different types of technologies and all the aspects which are associated with them. I found the idea of the **forum** especially **strong** because it allows people to discuss their views on the technologies.*
- *I found the theoretical information of the course: **good structured**, **interesting** and quite **complete**. The **forums are useful** to discuss the topics as well as the quiz to evaluate some ideas learned.*

It is not easy to find results from similar e-learning initiatives for a constructive comparison; nevertheless, the evaluation philosophy includes the same main topics than those used in previous experiences like Bacelar-Nicolau et al. (2009).

The interactive visual elements are a highly appreciated part of e-learning (Violante and Vezzetti, 2013). For the case presented was concluded than one of the options to be incorporated in future editions is the interactive videos, which is a very valued element by learners and contributes to a better learning performance (Gonzalez et al., 2010). On the other hand, it is known that, the level of satisfaction experimented by the students is an indication of the effectiveness in e-learning (Levy, 2007).

Editions made

From March to July 2010 five editions of the e-learning course (two weeks duration) were offered over the internet as part of the *ProDes* project training activities. Although the *ProDes* project concluded on October 2010, due to the great acceptance shown by students and the increasing interest registered, two more editions were offered out of project (three weeks duration).

After seven editions offered in English; the course contents were translated to French, Spanish and Portuguese. Two closed editions were offered in French for university students of Morocco, within the *TAKATONA* and *MOBADARA* projects. Also, a double closed edition in Spanish and Portuguese was offered to people from Canary Islands (Spain) and Cape Verde within the *ISLHáGUA* project.

Recently an improved and extended edition of the course was implemented (in English) co-organized together with the European Desalination Society (EDS). Finally, an on-going closed edition within the *RENOW* project is offered in French to 50 university students from Dakar (Senegal) and Nouakchott (Mauritania).

Summarizing, the main results obtained in the e-learning course along the **12 editions** made; we have that **489 students** from **35** different **countries** of the **five continents** have participated in the course (figure 7), which is a very relevant amount and shows the international impact of this initiative.

Figure 7: Student's distribution around the world.

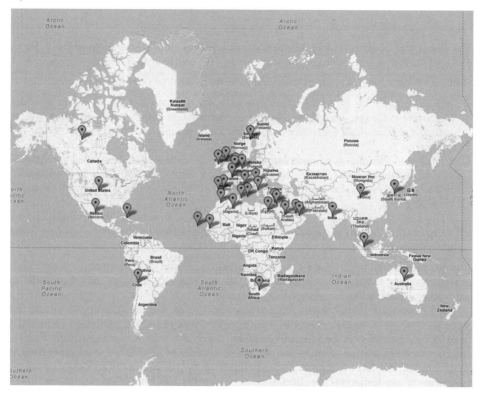

Figure 8: Student's origin in the 12 editions made

DES-RES course students origin

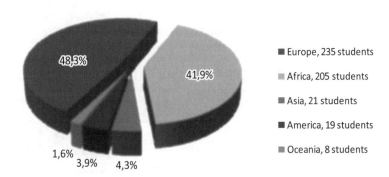

- Europe, 235 students
- Africa, 205 students
- Asia, 21 students
- America, 19 students
- Oceania, 8 students

As shown in figure 8 (see above), a great percentage of students (more than 48%) were from EU; this was quite expected due to the dissemination in the first seven editions were mainly made in the countries represented in the ProDes project partnership. Concerning the rest of students, it is remarkable the significant participation of African people (almost 42%).

Regarding the students professional occupation, it was different along the twelve editions. Except for the e-learning course closed editions, usually aimed at university students; in the rest of editions, professionals related with water or energy sectors were the collective with the most important participation, followed by students and researchers.

Improved and extended edition co-organized with the European Desalination Society (EDS)

An improved (more contents) and extended (1-month duration) version of the e-learning course has already concluded last 6th October in collaboration with the EDS (*http://www.edsoc.com*); see the course brochure in figures 9 and 10 (below).

In this edition contents were updated and completed with two new chapters, and an extra practical case was offered (solar thermal energy driven Multi Effect Distillation system).

Figure 9: Last course edition brochure (front view)

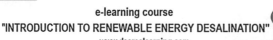

Figure 10: Last course edition brochure (back view)

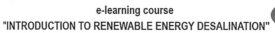

Also the Moodle platform version was updated to the last version available (figure 11).

Figure 11: View of the improved and extended course edition made with the updated Moodle version.

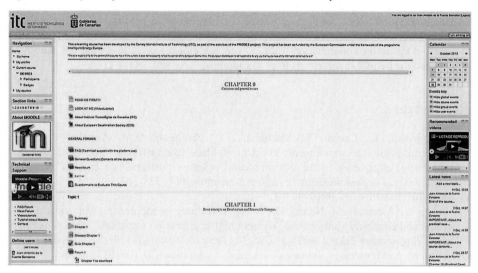

Conclusions

The main purpose of this e-learning course is to provide experts, professionals and postgraduate students from any country with the latest knowledge of the different existing systems and technologies on RE powered desalination. More specifically, the course presents the basic theory of desalination using RE. The experiences acquired so far, the current state of the art and the most promising initiatives.

Based on the impressions made by the students who have completed the course, we consider that the main objective of the course is fulfilled by far. After completing the course, students acquire an important knowledge about the fundamentals of RE– powered desalination.

According to the results obtained along the twelve editions made of the e-learning course we conclude that the experience has been very positive, not only for the students, but also for the teachers. Students have pointed out that the course is an opportunity of receiving interesting and diverse information in a very short period of time and of exchanging different points of view. On the other hand, the queries and comments of the students have contributed to enrich the contents of the platform.

The development of all these editions has allowed the authors to identify elements to be improved in future editions, as the visual elements.

Further information available at:
http://www.itccanarias.org/web/servicios/agua/formacion-eng.html;
www.desreslearning.com.

Acknowledgments

The authors want to express their gratitude to the entities involved in the PRODES project: the European Commission for their economic contribution to co-fund the project through the IEEE Program and the partners for their technical contributions. On the other hand, we want to point out the active collaboration from EDS to implement the editions after the end of the project, and the helpful involvement from the administrative staff of the ITC.

References

Bacelar-Nicolau, P., Caeiro, S., Martinho, A.P., Azeiteiro, U.M. Amador, F. (2009) "E-learning for the environment: The Universidade Aberta (Portuguese Open Distance University) experience in the environmental sciences post-graduate courses", *International Journal of Sustainability in Higher Education* 10(4),354-367 DOI: 10.1108/1467637 0910990701.

Benchicou, S., Aichouni, M. Nehari, D. (2010) "E-learning in engineering education: a theoretical and empirical study of the Algerian higher education institution" *European Journal of Engineering Education* 35(3), 325-342 DOI:10.1080/03043797.2010.483610.

González, M. J., Montero, E., Beltrán de Heredia, A., Martínez, D. (2010) "Integrating digital video resources in teaching e-learning engineering courses". IEEE EDUCON Education Engineering 2010 – The future of global learning engineering education. April 14-16, 2010 Madrid (Spain): 1789-1793 (available at *http://www.ieec.uned.es/Investigacion/ Educon2010/SearchTool/EDUCON2010/papers/2010S10E04 pdf*).

Levy, Y. (2007). Comparing dropouts and persistence in e-learning courses. *Computers & Education* 48(2), 185-204. DOI: 10.1016/j.compedu.2004.12.004.

Violante, M. G. and Vezzetti, E. (2013), Virtual interactive e-learning application: An evaluation of the student satisfaction. Comput. Appl. Eng. Educ. doi: 10.1002/cae.21580.

Education, Digital Inclusion and Sustainable Online Communities

Luísa Aires[1], Paulo Dias[1], José Azevedo[2], M. Ángeles Rebollo[3] and Rafael García-Pérez[3]

Abstract

Over the last decade, significant contributions have been made to a better understanding of sustainability and education. However, very little has been researched on sustainable education in the digital society. In new digital environments, collaborative practices, hybrid contexts, distributed knowledge or shared responsibility give rise to new approaches to the relationship between education, participation and digital inclusion – central constructs of sustainable education.

In this chapter, we propose a theoretical framework to analyze the relations between online communities, participation and digital inclusion. Sustainable education is interpreted as a shared and interactive process and is associated to community development across collaborative practices. In this framework, sustainable online communities embody the change of educational cultures.

The chapter is divided into four main parts. First, it provides a theoretical analysis of a collaborative and networking culture that underlies sustainable online communities. Second, it examines digital participation and literacy with regard to access and use, and as a means of participation equity and empowerment in social and educational practices. Third, the relationship between digital inclusion and gender in social networks is explored, taking into account some results from a research project on this topic. The chapter concludes by connecting digital inclusion and empowering practices with a view to developing sustainable online education.

Introduction

Sustainability and education has become an expanding area of research. As Hopkins and McKeown stated in 1999, education is an essential means to achieve a sustainable future; in other words, the path to a sustainable future begins with greater access to education. However, much of the research has focused on *education for sustainable development* or *education for sustainability* and not *sustainable education* and its paradigms, purposes and practices or its adequacy to address the challenges of a digital society (Sterling, 2008).

As noted by Sterling (2008:63), sustainable education means

> "a change of educational culture, one which develops and embodies the theory and practice of sustainability in a way which is critically aware. It is therefore a transformative paradigm which values, sustains and realizes human potential *in relation to* the need to attain and sustain social, economic and ecological well-being, recognizing that they must be part of the same dynamic" (Sterling, 2001:22).

---

1   Universidade Aberta (Portugal).
2   University of Porto (Portugal).
3   University of Seville (Spain).

Unlike *education for sustainable development* or *education for sustainability* that are focused on "how do we educate for sustainable development" (Sterling, 2008:63), sustainable education calls for an *anticipative* education that recognizes the new conditions, liabilities and discontinuities of global changes and new strategies and practices to address those changes *(idem)*. Following Sterling, sustainable education entails the following main descriptors: 1) sustaining – it sustains people and communities; 2) tenable – it is ethically defensible; 3) healthy – it is itself a viable system, nurturing healthy relationships; 4) durable – it works well enough in practice (Sterling, 2008:65). Therefore, as we will argue in this study, sustainable education is not just an "add-on" of sustainability concepts to parts of the educational curriculum. On the contrary, it means a cultural change in current educational models *(idem)*. This shift in perspective has a particular impact on the analysis of collaborative and participative processes in online education.

To educate in a changing, mediated and mediatized world (Couldry, 2008), strongly marked by the presence of digital and networked technologies, necessarily requires different forms of action (Weller and Anderson, 2013). But what kind of theoretical premises and courses of action can ensure a sustainable education in this changing context? The emerging digital cultures have significant implications in the mission of higher education institutions, as social transformers, in the fulfilment of the promise of the right to universal digital participation (Caswell et al., 2008). This new educational commitment requires the transformation of traditional educational conceptions, strongly influenced by formal education contexts. Hybrid forms should also emerge in which the formal and the informal are mixed in new online educational paths, inseparable from open communication and networking practices.

The participation ecologies of the digital society take into account the existence of open and hybrids systems, which are not compatible with a linear logic of knowledge construction. Thus, the sustainability of online education is part of a public pedagogy, a consequence of social networking, shared leadership communities, interpersonal and technological mediations, and formal and informal learning experiences (Dias, 2012). Therefore, it implies sustainability mediated by multiple flows and targeted at social inclusion. This perspective advocates a conceptual framework in which digital technologies are appropriated by agents embedded in communities whose leadership strategies are founded on interaction, sharing and trust (Dias, 2013a; Dias, 2013b). Sustainability is thus closely related to the construction and regulation processes of networks of agents organized into communities that are involved in the development of new identities and innovation processes.

The rapid expansion of online social networking has led to changes in the traditional roles of individuals, from passive consumers to agents who transform and produce knowledge. The sharing of online content on websites, blogs, Facebook or Twitter was the beginning of a practice that currently goes beyond the merely informational dimension. It has become a digital *agora* where the dynamics of *self* intertwine with social processes experienced by groups of different ages, socioeconomic standings or levels of schooling. The social network Facebook in particular has today become the setting in which a diversity of communities have developed – exchanging information, constructing and sharing knowledge, developing values and practices of citizenship and social inclusion, as well as cultural, educational or business activities.

This last topic completes the outline of the theoretical framework we propose to develop in this chapter, whose main goal is to analyze the processes of digital inclusion and participation in online communities, as structural dimensions of sustainable online education.

The chapter is divided into four main sections. After introducing the contribution of the participation culture and digital communication ecologies to sustainable education, a theoretical perspective is put forward to develop sustainable online learning communities. Then, digital inclusion and digital literacy are analyzed as central constructs in formal and informal participatory learning. The relationship between digital inclusion and gender in social networks is then explored, taking into account some results from a research project on this topic. The chapter concludes by connecting participation, collaboration and empowering practices in the development of sustainable online education.

The Sustainability of Online Communities

The educational sustainability of communities created in social networks can be traced back to Dewey, Bruner, Vygotski and Cole, among others. Dewey and Vygotsky argue that individual and society are not separate entities but two dimensions of a whole, and put forward an interpretation of human development based on interactions between individuals, contexts and cultural instruments.

As stated by Bruner (1960), learning combines not only concepts, categories or problem-solving strategies, but also the ability to use and reinvent things, giving them a personal trademark. Following this perspective, Cole (1996) argues that technologies or cultural artifacts are produced and modified throughout history in accordance with the purposes of their use. These artifacts may be material or ideal. Following Wartfosky's viewpoint, Cole proposes three levels to analyze cultural artifacts: primary artifacts – refrigerator, radio, television, computer; secondary artifacts – beliefs and norms that enable the preservation and transmission of modes of action; tertiary artifacts – artworks and their processes of perception.

If we focus on social networks like Facebook, we can see that, more than an artifact, these online networks appear as social and cultural laboratories where we can find different types of interaction and mediation, or even identify elements and dynamics of community regulation and sustainability that could have an encouraging impact on educational methods. In online social networks, the intentional processes that organize and develop the interactions among the community's members may anchor the innovation and sustainability of open education and networking. These processes are developed through conversational practices that expand from informal settings into the sphere of formal education, and may indeed enrich the latter. According to Dias (2013b), citing Downes "(...) the community that forms around the courses or subjects is a lot more important than the content." (2012). Several studies focusing on social networks as personal learning environments (Adell and Castañeda, 2010; Sloep and Berlanga, 2011) are of great value in creating synergies between formal and informal settings for learning and to adapt educational strategies to promote adult learning and digital inclusion. Those studies are supported by: a) a dynamic and comprehensive notion of learning as a process that encompasses different systems and contexts of activity, which provide people with opportunities to learn, and also involve adaptation and change to be permanently updated throughout life and, b) a perspective of the learner as an active agent who is able to decide what, how, when and where to learn, and above all, with whom to learn. Godfrey and Johnson (2009) propose a model to improve access and use of information by older people on the Internet based on the enrichment of digital circles of support; that is, this model is rooted in creating sustainable digital communities which complement and strengthen local communities where people take more active roles towards digital inclusion (Figure 1).

Learning, as a process emerging from a relational network of ideas, conceptions and representations, is based on this social process of building experiences by the community. Due to its complex nature, this process is not limited to the transfer of information but rather expands with the construction of knowledge. As mentioned by Brown and Adler, cited in Dias (2013b), "The focus is not so much on what we are learning but on how we are learning." (2008:18). To think of pedagogy of change – directed at the design and development of emerging learning environments – is an increasingly important issue, especially for open and networked education. In this context, educational sustainability is interpreted as a process built and regulated by the community, in the interactions and mediations that it promotes and is also associated to collaborative practices and to community development (Dias, 2013b).

Focusing this approach on Facebook, some researchers, such as Wen Tian and colleagues (2011), have stressed that virtual social networks not only have a leading role in the identity construction of users and groups, but also influence positively the social learning of students, as well as their academic trajectories. This scenario is a great challenge to educational institutions, to design flexible pedagogical models and focus on spiral curriculum development (Bruner, 1960), founded on the constructs shown in Figure 1. More important than the content itself is the ability to find different ways to solve problems and to encourage knowledge innovation.

Figure 1: Sustainability of online learning communities in social networks

(Adapted from Dias, 2013b)

The learning community is interpreted as a network of agents that mediates knowledge construction by collaborative and dialogical practices. These communities are organized through shared dynamic leaderships using available artifacts in open and hybrid contexts and promote a distributed knowledge. Open Educational Resources (UNESCO, 2011), as well as other instruments designed specifically for educational purposes, are significant mediational means in these learning communities. Thus, the boundaries between categories, such as formal and informal or global and local, are weak and result in broader frameworks of digital and social inclusion through particular forms of participation. The greater or lesser degree of innovation and integration of these practice strategies will promote a greater or lesser sustainability of these communities.

Digital Inclusion and Participation

The approach we propose for the analysis of online learning communities constitutes an intentional shift in pedagogical thinking. This perspective intends to promote a new pedagogy that enhances the value of interaction in a network of agents, the mediating role of contexts and artifacts in learning and reaffirms the role of a public pedagogy as a means to educate for participation and digital citizenship. This means there is a need for a shift to a more complex formulation that is able to establish a new research agenda. Three overlapping dimensions seem to be involved in pursuing such an objective – 1. the 'technological envelope', 2. the 'literacy envelope' and 3. the 'social participation envelope' of digital inclusion.

1. A more differentiated notion of access – a need to further our understanding of how the choices that people make regarding the nature and extent of their use of technology may be influenced by technological factors and, especially, the availability of social support to deal with problems that predictably occur with information technology (*the technological envelope*).

   Accordingly to Seale (2009), conceptualizing digital inclusion as involving access to technologies also raises important issues relating to technological determinism, the digital divide and the complexities of bringing about change to mitigate inequalities. The data from digital divide surveys has tended to be interpreted rather simplistically, with assumptions being drawn that all that is required is the provision of technology and training on how to use it. Many digital divide researchers have however warned that digital exclusion is a complex phenomenon that needs to be understood in broader terms than just access to equipment and use (Hargittai, 2010). It has to include social and cultural factors that lead many individuals and groups to choose exclusion, in what may be called the digital choice versus the digital divide (Dutton et al., 2007). Empowerment is also increasingly being understood as individuals exerting control and choice by making decisions about appropriate or meaningful technology use. For example, Selwyn (2004), on the low use of computers, identifies a hierarchy of engagement with technology, ranging from absolute non-users, to lapsed users and rare users.

2. More differentiated measures of digital literacy that include a diversity of uses and skills (*the literacy envelope*). According to Hague and Williamson (2009), digital literacy means knowing how technology and media affect the ways in which we go about finding things out, communicate with each another, and gain knowledge and understanding. And it also means understanding how technologies and media can shape and influence the ways in which school subjects can be taught and learnt. In a dense landscape of information sources, communication opportunities and tools for creating new digital objects, teaching and learning cannot be confined to pen and paper activities. The implications of this perspective for educational contexts are summarized by Street:

   > 'the literacy demands of the curriculum [involve] a variety of communicative practices, including genres, fields, and disciplines. From a student's point of view a dominant feature of academic literacy

practices is the requirement to switch practices between one setting and another, to deploy a reper-
toire of linguistic practices appropriate to each setting, and to handle the social meanings and identities that each evokes.' (Street, 2004:15).

According to Goodfellow (2011), scholars of 'multiliteracies' argue that it is not useful to
think of literacy education solely in terms of developing generic competences that can be
transferred from context to context. Rather, what several authors (Lankshear and Knobel,
2008 and others) refer to as 'digital literacies' are not simply the skills involved in using
digital communication, but the diverse ways of making meaning that involve 'digital en-
codification', and the 'enculturations that lead to becoming proficient in them'.

3. Closer examination of Internet use as contextualized by education. This means examining
    how use fits into people' lives, on the one hand, and how education styles and practices
    shape Internet use (*the social participation envelope*), on the other. The ways of pursuing
    participation in society have become ample. Modern e-participation seeks to empower
    people with the help of Information and Communication Technologies (ICTs). Modern
    ICTs offer learners more possibilities of when, where, how, and by which means they
    want to have educational experiences (DiMaggio et al., 2004; Livingstone and Helsper,
    2007; Hargittai and Hinnant, 2008; van Dijk, 2005; Hargittai, 2010). Accordingly to
    Fuchs and Obrist (2010), participation is a key-element for sustainability in the current in-
    formation society. But this does not only leave people with a variety of participatory
    means at their hands, it raises major challenges for them too. We argue that participation
    involves making informed choices about technology use. If people want to move from
    occasional e-participation to sustainable e-participation, they need to learn how to make
    informed choices among the available ICTs and use them appropriately. According to
    Jenkins et al. (2006) they may establish new affiliations by becoming members of formal
    and informal online communities. As ICTs became more and more interactive, their adop-
    tion by society basically provides for a more participatory culture. Current digital inclu-
    sion research has failed to produce a detailed review of what constitutes support for par-
    ticipation from educational institutions. A lack of open and reflexive accounts of practice
    means that we are no closer to identifying and understanding the kinds of participation
    practices that are required to challenge digital exclusion.

Digital Inclusion and Learning:
Women in Online Social Networks

Many studies have shown gender differences in uses and relationships between people and
technologies (Wasserman and Richmond-Abbott, 2005; Hilbert, 2011). However, the Web
2.0 and, especially, social networking have changed the presence and participation of women
in virtual environments. Some studies reveal that social networking activities related to maintaining contact with friends and family are led mainly by women (Clipson et al., 2010; Mazman and Usluel, 2011).

What do social networks do to women's lives? Why are they more meaningful to them
than other technologies? What experiences have value for them and to what extent can they
use them to change / improve their lives? In a previous study on the use of social networks by
adult women, Rebollo, García and Sanchez-Franco (2013) found that the main motivation for
using social networks was relational. In this study, the data on women indicate that social
support and the quality of relationships in social networks are important aspects of their on-

line learning (see also Oblid Network). The results show that 69.4% of the women surveyed perceived a medium-high level of support, whereas only 30.6% find little or no support in social networks. Women pointed out five significant aspects of support on social networks: count on people with whom you can enjoy yourself, count on people who express affection for you, who you can talk to, share interests with and trust. Women also believe that the quality of their relationships on social networks is associated with satisfaction and trust. The study found that about 80% of women get a medium-high level of satisfaction and trust in the relationships they establish on social networks.

Women also expressed the feeling of being socially integrated in a community or group in their interviews. Their narratives also reveal experiences and relationships over time and space, as well as actively participating in the cultural life of local communities despite geographical distance:

(M):   *The classmates who you lost contact with twenty-five years ago, (...) thanks to Facebook I got back in touch with them and I've been reunited with people, as I've already told you, after twenty-five or twenty-six years of not hearing from them or knowing anything about them, and now they are my best friends.*

(D):   *Well, it's that, to communicate with, to communicate with my sons, old friends I've regained (...) and then ... it is very nice because you can see people again who had disappeared from your life, and also, there's a lot of curiosity, not simply Facebook, the Internet gives me the opportunity to go to thousands of pages (...) enter many websites, visit museums, see cities, ... I mean, there are so many things.*

(A):   *It provides me with everything, I keep in touch with lots of people, I've been able to relate to my environment, because I'm not from here, I'm from a town of Huelva and Facebook has allowed me to live everything from there through my people.*

(I):   *I connect a lot with an association because of an illness that my daughter has and it's very good for me because when you're diagnosed with a disease and it hits you all at once, doctors may be able to explain it you... but you need more, don't you? (...) And besides I can also connect with many people, for example, I also paint and we send each other paintings, "I'm doing this," "I'm doing that," "and I tried this technique using acrylics", "and I the other'.*

Furthermore, the women in this study claim to have learned, in informal settings, about very specific needs and they have done so with the support of certain people in their personal network. The data reveal that social networks can become personal learning environments because they allow you to keep in touch with many different people (age, gender, race, social class, etc.) with whom you share interests and in whom you can trust; in other words, you can develop a network of mutual care and support. In the following excerpts, the women express how their learning took place on the use of online social networks.

(D):   *I'm very curious, I get into everything... well, I started with the computer many years ago and, of course, my youngest daughter taught me some things as well. Hmm, then the City Hall gave a course for older people and I started to learn more, not much, because you have this curiosity and, of course, you make thousands of mistakes, you mess things up thousands of times*
          ...

(Lo):   *No, I haven't taken any courses. This comes from working hard, getting into things, that I mu::st:, (...) and most of all it's because of my daughter, since she began, she's the one who says "mom you cannot go there, mom you should click here". My daughter's been so patient, she told me "mom here, mom there". If I want to look for a recipe, "mom, you have to click on this small icon to get it, mom, you have to click here to go back."*

However, this study finds unmet needs related to learning and discovering new dimensions of their identity that would enable social networks to eventually become environments of growth and empowerment. In the next excerpts we can find some of their concerns about the uses of social networks and the role they can play in their lives:

(L):   *It's also a risky tool because once someone wrote on Facebook, "Facebook brings people who are far away closer, but it also distances people who are nearby." I think this is so true. We begin to contact other people and we forget about those who are here. That's also dangerous.*

(El): *When do I have time, huh? I mean not always ... but sometimes when I've finished, when I start cooking, and you know, when all of the chores are done, and I say, ok now let's look at this a little, and that's how I'm doing it.*

(T):   *I have my limits. I really, really don't like all the gossip ... but I want to keep my privacy because I have a profile, but I haven't filled it in, all the fields are empty, you know? Female... and nothing else, that's it, I haven't even put my age, we don't use it properly, sometimes it's used to distract, but telling lies or ... stories .. and (...) I don't know what it's all about ... I would never post anything about my real life.*

As we have seen in these excerpts, some of the concerns expressed by women using social networks refer to difficulties in reconciling their online life and their daily life (activities outside and inside the network), the use of time, the feeling of loss of control over certain aspects of their private life and the blurry boundaries between public and private on social networks.

This study also highlighted that there is a need to measure/examine indicators on the quality of relationships regarding the regulatory mechanisms of interactions, that is, ways of reciprocal support which are used for the shared construction of knowledge and experience in social networks. On the other hand, it is important to consider that not only individuals but also communities must be efficient in the use of resources, involving the development of effective and planned forms of communication on social networks (Figure 1).

According to Sloep and Berlanga (2011), the way we have organized early education in our societies makes it inappropriate to educate adults to acquire specific competencies needed in their lives. The knowledge society, characterized by the relentless production of knowledge, has changed the ways in which we learn and has broadened the learning contexts. Virtual learning communities are groups of interest or participation that use digital resources in two ways: as an infrastructure to consolidate and expand networks of communication, and as a tool to promote and enhance learning (Coll et al., 2007). Besides sharing very similar interests, a learning network provides participants with resources and services to meet their particular goals (Sloep and Berlanga, 2011). Furthermore, Koper (2009) argues that, in their efforts to acquire skills in a learning network, people can exchange experience and knowledge, collaborate on projects, create working groups, offer and receive support from others or assess each other.

Conclusions: Digital Inclusion and Empowering Practice for Sustainability

While, in principle, it is relatively easy to accept the argument that supportive organizations and practitioners are necessary to enable learners to make empowered choices, it is more difficult to create these resources by defining and describing exactly the kind of supportive practices that are the most effective or successful in terms of promoting the digital inclusion of learners.

A participatory society, which involves its members on a large scale, requires educated and skilled citizens. Skills, which we prefer to call media/ICT literacy, include the ability to access, navigate, appraise, and create content by means of information and communication technologies (Mansell, 2009:108). From a critical perspective, Mansell argues that people are increasingly dependent on such literacy and we must therefore bear in mind that media skills and literacy are not equally distributed within society. We argue that skills and competencies are developed comprehensively and are socially balanced, if supported by means of participatory culture offers. Participatory culture, at its best, represents an ideal learning environment where sustainability emerges from collaborative practices within online communities. Higher education research and practice has been slow to embrace the cultural impact of the new communication order in its literacy practices, within a framework of sustainable education. Moreover, expanding the concept of literacy to include communication in general may have begun to blur the distinctions between practice in the school and post-school sectors, so that discussion of literacies of the digital kind, usually rooted in schools, is now assumed to be applicable across sectors. This transformative paradigm provides a new educational ecology, towards sustainable education in the digital society (Sterling, 2011). While formal learning usually occurs in schools, informal participatory learning takes place in affinity spaces, both offline and online in cyberspace. New media practices are producing a 'profound shift' in educational culture in a social as well as a pedagogical sense, serving as powerful opportunities for learning because *"they are sustained by common endeavors that bridge differences in age, class, race, gender, and educational level, and because people can participate in various ways according to their skills and interests [...]"* (Jenkins, 2006:9).

References

Adell, J. & Castañeda, L. (2010). Los Entornos Personales de Aprendizaje (PLEs): una nueva manera de entender el aprendizaje. In R. Roig Vila & M. Fiorucci (Eds.), *Claves para la investigación en innovación y calidad educativas*. Alcoy: Marfil – Roma TRE Universita Degli Studi.

Bruner, J. (1960). The process of education. Cambridge, Mass: Harvard University Press.

Caswell, T.; Henson, S.; Jensen, M.; Wiley, D. (2008). "Open Educational Resources: Enabling universal education". *International Review of Research in Open and Distance Learning*, 9(1), ISSN: 1492-3831.

Clipson, T. W., Wilson, A. & DuFrene, D. (2012). The Social Networking Arena: Battle of the Sexes. *Business Communication Quarterly*, 75(1), 64-67.

Cole, M. (1998). Cultural Psychology. A once and a future. Cambridge, Mass: Harvard University Press.

Coll, C.; Bustos, A. & Engel, A. (2007). Configuración y evolución de la comunidad virtual MIPE/DIPE: retos y dificultades. *Revista Electrónica de Teoría de la Educación*, 8 (3), 86-104.

Couldry, N. (2008). "Mediatization or mediation? Alternative understandings of the emergent space of digital storytelling". *New media & Society*, 10 (3), 373-391 [DOI: 10.1177/ 14614448 08089414].

Dewey, J. (2004). Educación y Democracia. Madrid: Ediciones Morata.

Dias, P. (2012). Comunidades de educação e inovação na sociedade digital. *Educação, Formação & Tecnologias*, 5 (2), 4-10 [Online], Available at: *http://eft.educom.pt*.

Dias, P. (2013a). Inovação na educação aberta e em rede, VIII Conferência Internacional de TIC na educação – Challenges 2013. Braga: Universidade do Minho. Available at: *http://www.aca demia.edu/4082025/Inovacao_na_educacao_aberta_e_em_rede*. Accessed: September 2013.

Dias, P. (2013b). Inovação pedagógica na educação aberta e em rede. Revista *Educação, Formação e Tecnologias* (in the press).

DiMaggio, P., Hargittai, E., Celeste, C., & Shafer, S. (2004). Digital inequality: From unequal access to differentiated use. In K. Neckerman (Ed.), *Social Inequality* (pp. 355-400). New York: Russell Sage Foundation.

Dutton, W. & Helsper, E. (2007). Oxford Internet Survey 2007 Report: The Internet in Britain. Oxford: Oxford Internet Institute.

Fuchs, C. & Obrist, M. (2010). "Design principles for a Participatory, Co-operative, Sustainable Information Society (PCSIS)". In *HCI and society: Towards a typology of universal design principles. International Journal of Human-Computer Interaction*, 26(6), 638-656.

Godfrey, M. & Owen, J. (2009). Digital circles of support: Meeting the information needs of older people. *Computer in Human Behavior*, 25, 633-642.

Goodfellow, R. (2011). "Literacy, literacies, and the digital in higher education". *Teaching in Higher Education*, 16 (1), 131-144.

Godfrey, M. &, Johnson, O. (2009). Digital circles of support: Meeting the information needs of older people. *Computer in Human Behavior*, 25, 633-642.

Greenhow, C. & Robelia, B. (2009). "Old Communication, New Literacies: Social Network Sites as Social Learning Resources". *Journal of Computer-Mediated Communication*, 14(4), 1130-1161.

Hague, C. & Williamson, B. (2009). Digital participation, digital literacy, and school subjects: a review of the policies, literature and evidence. Available at: *http://www.futurelab. org.uk/re sources/documents/lit_reviews/DigitalParticipation.pdf.*

Hargittai, E. & Hinnant, A. (2008). Digital Inequality: Differences in Young Adults' Use of the Internet. *Communication Research*, 35, 602-21.

Hargittai, E. (2010). Digital Na(t)ives? Variation in Internet Skills and Uses among Members of the "Net Generation". *Sociological Inquiry*, *80* (1), 92-113.

Hilbert, M. (2011). Digital gender divide or technologically empowered women in developing countries? A typical case of lies, damned lies, and statistics. *Women's Studies International Forum*, 34, 479-489.

Hopkins, C. & McKeown, R. (1999). Education for sustainable development". *Forum for Applied Research and Public Policy*, 14(4). 25.

Jenkins, H.; Clinton, K.; Purushotma, R.; Robison, A. & Weigel, M. (2006). Confronting the Challenges of Participatory Culture: Media Education for the 21st Century. Available at: *http://www.newmedialiteracies.org/files/working/NMLWhitePaper.pdf*; Retrieved July 14, 2010.

Koper, R. (2009). Learning Network services for professional development. Berlin: Heidelberg Springer.

Lankshear, C. & Knobel, M. (2008) (eds.). Digital literacies: Concepts, policies and practices. New York: Peter Lang.

Livingstone, S., & Helsper, E. (2007). Gradations in digital inclusion: Children, young people, and the digital divide. *New Media and Society*, 9, 671-696.

Mansell, R.; Avgerou, C.; Quah, D. & Silverstone, R. (2009). The Oxford Handbook of Information and Communication Technologies. Oxford: Oxford University Press.

Mazman, S.G. & Usluel, Y.K. (2011). Gender differences in using social networks. *TOJET: The Turkish Online Journal of Educational Technology*, 10(2), 133-139.

ObLID Network. Available at: *www.contemcom.org.*

Rebollo, M.A.; García, R. & Sánchez-Franco, M. (2013). La inclusión digital de las mujeres en las redes sociales online: un estudio en mujeres de zonas rurales de Sevilla [Digital inclusion of women on social networks: a study on rural women of Seville (Spain)]. Sevilla: Diputación Provincial de Sevilla.

Seale, J. (2009). Digital. Inclusion: a Research Briefing by Technology Enhanced Learning. Phase of the Teaching and. Learning Research Programme. University of Southampton. Available at: *http://www.tlrp.org/docs/DigitalInclusion.pdf.*

Selwyn, N. (2004). Reconsidering political and popular understandings of the digital divide. *New Media & Society*, 6, 341-362.

Sloep, P. & Berlanga, A. (2011). Learning Networks, Networked Learning. *Comunicar*, 37, 55-64.

Street, B.V. (2004). Academic Literacies and the 'New Orders': Implications for research and practice in student writing in HE'. *Learning and Teaching in the Social Sciences* 1(1), 9-32.

Sterling, S. (2008) 'Sustainable education – towards a deep learning response to unsustainability', Policy & Practice: A Development Education Review, Vol. 6, Spring, pp. 63-68. Available at: *http://www.developmenteducationreview.com/issue6-perspec tives1*; Accessed: January 2014.

UNESCO (2011). Guidelines for open educational resources (OER) in higher education. Available at: *http://www.unesco.org/new/en/communication-and-information/resources/ publications-and-communication-materials/publications/full-list/guidelines-for-open-educational-resources-oer-in-higher-education/.* Retrieved October, 2013.

Van Dijk, J. (2005). The deepening divide: inequality in the information society. Thousand Oaks, CA: Sage Pub.

Vygotsky, L. (1987). Mind in Society. The Development of higher psychological process. Harvard University Press.

Wasserman, I. M. & Richmond-Abbott, M. (2005). Gender and the Internet: Causes of Variation in Access, Level, and Scope of Use. *Social Science Quarterly*, 86, 252-70.

Wen Tian, S.; Yan Yu, A.; Vogel, D. & Chi-Wai Kwok, R. (2011). The impact of online social networking on learning: a social integration perspective. *International Journal of Networking and Virtual Organizations*, 2011; 8 (3/4): 264-280 DOI: 10.1504/IJNVO. 2011.039999.

Weller, M.; Anderson, T. (2013). "Digital resilience in higher education". *European Journal of Open, Distance and e-Learning*, 16(1) p. 53.

E-learning for sustainable development: the way ahead

Walter Leal Filho[1]

Abstract

This short paper outlines some of the areas where action is needed, so as to allow a more systematic development of e-learning as a tool for education for sustainable development.

Introduction: the contribution of e-learning to education for sustainable development

It is widely acknowledged that e-learning can be used as an effective tool to support learning on, through and about the environment. Some previous works have emphasised the need for and the relevance of innovative approaches to promote sustainability in higher education (e.g. Leal Filho, 2010a), whereas other works have illustrated what can still be done (e.g. Leal Filho, W. 2010b). In addition, the chapters presented in this book have documented a wide range of approaches, methods and techniques which show how e-learning applied to sustainable development may be realised.

In particular, if one considers that strategies to implement education for sustainable development should consider the surrounding environment and geographical locations, e-learning have a substantial contribution to provide.

Table 1 illustrates some of the many advantages of e-learning as a tool towards education for sustainable development.

Table 1: Advantages of e-learning

Item	Advantages
Approach	Action oriented
Focus	Diverse, with an on theoretical elements or practical considerations
Pedagogical nature	Inclusive, with the possibility to achieve pre-set goals
Placement in the curriculum	Flexible, suitable to fit even into pre-defined schedules and time-tables
Methodology	Use of e-media and tools may raise further interest
Students' participation	Actively encouraged with provisions for active learning
Availability of materials	Storage may make materials available to hundreds or thousands of students at the same time

One important component of the formula – and perhaps a starting point – is to ensure institutional commitment for the proposed e-learning efforts by looking at the current policies, visions and ethos of a given university, and providing a basis upon which e-learning can be implemented at the institutional level. This is unfortunately largely missing, and some recent

---

1   Hamburg University of Applied Sciences, Hamburg, Germany.

works on universities' commitment to sustainability (e.g. Lozano, Lukman, Lozano, Huisingh, Lambrechts 2013) have not specifically tackled this need.

Advantages of the use of e-learning on an education for sustainable development context

On the basis of the potential advantages offered by e-learning, the author defends the view that e-learning methods should become more popular and more widely used in teaching on education for sustainable development, in both higher education institutions and beyond. E-learning can also assist universities in becoming more international (Harris 2008), and to improve trends in work practice (Saeudy 2014). This is so based on two main elements:

1. The use of e-learning caters for students' engagement and participation;
2. One of the main barriers to the wide dissemination of education for sustainability, namely lack of resources, can be addressed by means of e-learning since it may cater for various materials (e.g. photos, films, texts, tables, etc) to be used in an integrated and flexible way, making learning processes more active and hence more interesting.

Further strategic advantages of using e-learning on education for sustainable development are:

❖ Flexibility;
❖ Cost-effectiveness;
❖ Learner centered dimension;
❖ Durability;
❖ Wide availability of materials suitable for use by large audiences.

But despite the potentials of e-learning, there are some problems which have been hindering progress. In particular, the fact that not all teachers are familiar with the diversity of e-learning methods and tools, means that many good opportunities to widen its use are currently being missed.

Conclusions: the way ahead

Since both the availability and range of ICT-based techniques and materials is rapidly evolving, the use of e-learning as a tool for education for sustainable development is likely to increase in the future. But in other to allow that full advantage of the potentials it offers can be taken, the way ahead needs to paved by a set of measures. These are:

a) an improvement in the amount, quality and attractiveness of pre-service and in-service teacher training, so as to foster the competencies needed among teachers to use e-learning in a meaningful way;
b) the targeted preparation of materials suitable for use on e-learning programmes, which may support teachers in delivering content which is relevant in support of sustainability education. In this context, manuals and lecture notes for training courses, to be used by teachers, need to be produced and more widely used;
c) the need to create synergies and links between extra-curricular and curricular learning, thus improving the base of knowledge of students on matters related to education for sustainable development.

Moreover, the documentation and dissemination of experiences and good practice, as this book has tried to do, is also a major need to be addressed.

Due to the innovative nature of e-learning and the inherent attractiveness of this field, secondary and higher education institutions should strive to make a better use of its potential, by focusing for example on themes such as climate change, renewable energy or biodiversity – among others- which are among the most important issues of modern times, and in relation to which improvements in respect of sustainability-based teaching and learning are urgently needed.

As this short paper has tried to demonstrate, e-learning can not only help to increase the capacity to understand the links between environment and sustainable development, but also to establish and develop a broader awareness of environment and social issues, going over and above what traditional learning methods can achieve.

References

Harris, S. (2008) Internationalising the University. Journal of Educational Philosophy and Theory, 40(2), pp. 346-357.

Leal Filho, W. (ed) (2010a) Sustainability at Universities: Opportunities, Challenges and Trends. Peter Lang Scientific Publishers, Frankfurt.

Leal Filho, W. (2010b) Teaching Sustainable Development at University Level: current trends and future needs. In *Journal of Baltic Sea Education* 9 (4), pp. 273-284.

Lozano, R., Lukman, R., Lozano, F., Huisingh, D., & Lambrechts, W. (2013). Declarations for sustainability in Higher Education: Becoming Better Leaders, Through Addressing the University System. *Journal of Cleaner Production*, Volume 48, pp. 10-19.

Saeudy, M. (2014) Sustainable Practices in the Academic Workplace. *The Journal of Academic Development*, Volume 1, pp. 29-40.

About the authors

Luísa Aires, Ph.D., Assistant Professor, Department of Education and Distance Learning, Universidade Aberta. Her research focuses on Communication, Digital Inclusion and Adult Education, privileging a sociocultural perspective. Her latest articles examine parents' and teachers' narratives about ICTs and daily life experiences.

luisa.aires@uab.pt

Rosa María Martín Aranda was educated at the Autónoma University of Madrid (Spain) from which she received her Master Science Degree in 1988, and her Doctorate (PhD) in Chemistry in 1992. She was Post-doc in the Instituto Superior Técnico de Lisboa (Portugal) during 1994. Since 1992, she teaches Inorganic Chemistry, Sustainable and Green Chemistry at Universidad Nacional de Educación a Distancia (UNED), Madrid, Spain. She is full professor of Inorganic Chemistry, Faculty of Sciences-UNED. She was Vice-Dean of Environmental Sciences (2006-2010). Since 2010, is Vice-Rector of evaluation procedures of UNED. She has been always involved in sustainability and efficiency projects.

rmartin@ccia.uned.es

Leanna Archambault, PhD, is an assistant professor of educational technology at the Mary Lou Fulton Teachers College at Arizona State University and the Lead Researcher/Liaison for the Sustainability Science Education Project. Through researching and teaching the Sustainability Science for Teachers course, Dr. Archambault seeks to have a lasting impact on the practice of future teachers throughout the state of Arizona. Dr. Archambault's research areas include teacher preparation for online and hybrid classrooms, as well as the nature of technological pedagogical content knowledge. Dr. Archambault has emerged as a leader in her field. In addition to publishing in several prominent journals, she was awarded the Online Learning Innovator Award for Important Research from the International Association for K-12 Online Learning in 2010 and 2012. Prior to taking her position at Arizona State University, Dr. Archambault graduated from the University of Nevada, Las Vegas with a PhD in instructional and curricular studies. As a former middle school English teacher, Dr. Archambault is passionate about improving education, particularly through the use of relevant and emerging technologies.

Leanna.Archambault@asu.edu

Ulisses Miranda Azeiteiro, was educated at the Universidade de Aveiro, Portugal, from which he received his Master Science Degree in 1994, Universidade de Coimbra, Portugal, where he received his Doctorate (PhD) in Ecology in 1999 and Universidade Aberta, Portugal, where he received his Aggregation in Biology in 2006. He holds an Assistant Professor teaching position at the Universidade Aberta (Assistant Professor with Aggregation), Portugal, in the

Department of Sciences and Technology, where he teaches Environmental Education, Education for Sustainable Development, Biology, Biodiversity and Research Methodologies. He is Senior Researcher at the Centre for Functional Ecology of Coimbra University where he is the Coordinator of the Group Ecology and Society.

ulisses.azeiteiro@uab.pt

José Azevedo, Ph.D., Associate Professor, Department of Sociology, University of Porto. Fulbright Scholar – University of Texas at Austin and Visiting Scholar at various international universities. His primary research interests are in Digital Divide, Public Understanding of Science and Science Documentary.

azevedo@letras.up.pt

Viraja P. Bhat was educated at the Karnataka Univeristy from where she received her Bachelor of Engineering degree in Electronics and Telecommunication in 1993, received her Master's in Business studies degree from Pune University in 2010. She is pursuing her Ph.D from Symbiosis International University in the subject E- Waste Management. She holds an Assistant Professor teaching position at Symbiosis Institute of International Business in the department of Information Technology, where she teaches Business Intelligence, Enterprise Resource Planning, Data Analysis and Project Management. She is the faculty mentor for the social initiative Kshitij under which she co-ordinates many projects for sustainability and waste management.

Christine Nasimiyu Bukania is the Knowledge Management Coordinator at the Cameroon-based Environmental Governance Institute (EGI). She holds a Master Degree in International Media Studies from the Deutsche Welle Academy in Bonn, a Masters in German and Cultural Studies from Nairobi and a Bachelor in Education from Kenyatta University in Kenya. She has over a decade of experience in education, communication and advocacy, especially in East Africa.

Sandra Sofia Caeiro holds an undergraduate degree in Environmental Engineering from Universidade Nova de Lisboa (1992), a Masters in Science of Coastal Zones from Universidade de Aveiro (1997) and a Doctorate on Environmental Engineering from the Universidade Nova de Lisboa (2004), Portugal. She is currently Assistant Professor in the Department of Science and Technology at Universidade Aberta (UAb) and a senior researcher at CENSE – Center for Sustainability and Environmental Research from Universidade Nova de Lisboa, Portugal and she also collaborates with the e-learning Laboratory of UAb. Her main research and teaching areas include environmental risk assessment, environmental management and spatial planning support tools, education for sustainable development. She is the coordinator of the B.Sc. programme on Environmental Sciences, and the institutional coordinator of the European Virtual Seminar in Sustainable Development. She is on the editorial board of the international journals of Ocean and Coastal Management, Elsevier, Latin American Journal of Management for Sustainable Development, Inderscience, and BioMed Research International, Hindawi Publishing Corporation and reviewer of several international scientific journals and books. She has mentored several post-graduate students and postdoctoral researchers, published papers in peer-review ISI journals, chapter books and international conference proceedings, and coordinated and participated in several research projects.

scaeiro@uab.pt

Fernando José Pires Caetano holds an undergraduate degree in Technological Chemistry from Faculdade de Ciências, Universidade de Lisboa and a degree in Master in Science and a Doctorate, both in Chemical Engineering from Instituto Superior Técnico, Universidade Técnica de Lisboa, at Portugal. Currently, he is an Assistant Professor at Universidade Aberta (UAb), Portugal, and a researcher at Centro de Química Estrutural (CQE) at Instituto Superior Técnico (currently Universidade de Lisboa). He is also a collaborator researcher at the e-learning Laboratory of UAb. He is the Director of Sciences and Technology Departement (Departamento de Ciências e Tecnologia – DCeT) of Universidade Aberta and was the coordinator of the B.Sc. programme on Environmental Sciences at UAb. His teaching subjects include Environmental Chemistry, New Energies, Environmental Technologies and Safety and Health Security at Work and the research activities have been mainly on Fluid Thermophysical Properties at CQE and Science teaching at UAb. He has mentored several post-graduate students, published papers in peer-review ISI journals, chapter books and has participated in several International Conferences and in several research projects.

Sally Caird is a Research Fellow in Design and Innovation in the Department of Engineering and Innovation, Faculty of Maths, Computing and Technology at the Open University. Dr Caird has research interests in environmental assessment methodologies, and the design and innovation processes associated with sustainability in complex higher education service-systems. Her research interests also incorporate the processes associated with design development, performance, use and diffusion of low carbon products and systems in the built environment. She was Investigator on the SusTEACH research and development project 2011-2013 funded under the UK Jisc Greening ICT programme (*http://www.open.ac.uk/blogs/susteach/*), which was shortlisted for the Green Gown Award 2012 (*http://sustain abilityexchange.ac.uk/research-and-development/1377-green-gown-awards-2012-research-a-develop ment-the-open-university-finalist-case-study-avideo*), and later developed as a new teaching unit for the OpenLearn platform *(http://www.open.edu/openlearn/nature-environ ment the-environment/the-environmental-im pact-teaching-and-learning/content-section-0).*

Sally.Caird@open.ac.uk

Amelia Clarke has been working on environment and sustainability issues since 1989, and has been recognized as an environmental leader in Canada. She is now an Assistant Professor in the School of Environment, Enterprise and Development (SEED) at the University of Waterloo and is Director of the Master of Environment and Business (MEB) executive-education online program. She teaches: Introduction to Environment and Business, Introduction to Sustainability for Business, and Strategies for Sustainable Enterprises. Her main research interests are: community sustainable development strategies; corporate social and environmental responsibility; campus environmental management; collaborative strategic management; cross-sector partnerships; and youth-led social entrepreneurship. Dr. Clarke sits on the editorial board of the Academy of Management Learning and Education (AMLE) journal, and is an executive member of the Social Responsibility Division of the Administrative Science Association of Canada (ASAC). She holds a PhD in Management from McGill University.

amelia.clarke@uwaterloo.ca

Ron Cörvers is director of ICIS, the International Centre for Integrated assessment and Sustainable development (*www.icis.unimaas.info/*), and associate professor governance and sustainable development at Maastricht University. He is education director of the Master program Sustainability Science and Policy (*www.maastricht university.nl/masterssp*) and managing director of the Maastricht-Utrecht-Nijmegen Program on Partnerships (*www.munpop.nl*). Before his current position he was associate professor environmental policy at the School of Science, Open University in the Netherlands. He holds a Master diploma environmental geography from Radboud University Nijmegen and a PhD degree from Utrecht University. His current research is on governance and sustainable development, global certifying partnerships, and learning for sustainable development.

Paulo Dias, Rector of Universidade Aberta, Portugal. He obtained his Ph.D. in Education in 1990, from the University of Minho, Portugal, and is full professor in Curriculum Development and Educational Technology. His main areas of research are Distance Education and Elearning; Open Education and Pedagogical Innovation; Collaborative and Social Learning in Digital Society, within which he coordinates and participates in national and European Community programmes.

paulo.dias@uab.pt

Gary Dishman is teaching project assistant at the School of Geography, Planning and Environmental Management at the University of Queensland, Australia.

Juan A. de la Fuente (Chem. Eng and Marine Sci. Grad.): Researcher of the Water Department. He is expert on the design, operation and optimization of RO desalination plants. He has been involved in assessment of wind and PV powered RO units and testing of different energy recovery systems for low capacity RO desalination plants since 2007. Teacher of Renewable Energy Desalination courses. 6 publications

Rafael García-Pérez, Ph.D., is Associate Professor in Techniques and Tools of Assessment of Educational Processes at the University of Seville (Spain). His main research area is focused on Assessment of Educational Processes in Virtual Environments. He is currently conducting research on the quality of rural women's relationships as an explanatory factor of their digital inclusion in social networks.

rafaelgarcia@us.es

Anthony Halog is a lecturer in industrial environmental management and a research group leader of sustainable consumption and production in the School of Geography, Planning and Environmental Management at the University of Queensland, Australia. His research focuses on the sustainability of the human-nature complexity through understanding the nexus of material and energy systems. Over the past years, he has been working on the theoretical foundation and industrial applications of Life Cycle Sustainability Assessment (LCSA),

Sustainable Operations Management, Corporate Environmental Management, and Greening Supply Chains. He is interested in the life cycle of manufactured goods and, ultimately, wastes – and in the environmental and economic potential of circular and green economy.

a.halog@uq.edu.au

Svetlana Ignatjeva, Dr. phys, is a docent at Daugavpils University, She is the head of Department of Computer Science. Her research interests are data processing and IT technologies. She conducts ITC courses in teacher training programs.

Dzintra Ilisko, PhD is an associate professor at Daugavpils University, Faculty of Education and Management. She is a head of the Institute of Sustainable Education at Daugavpils University. She is a member of international associations and networks such as SWEDESD (Swedish International Centre of Education for Sustainable Development) network; the BBCC consortium (Baltic & Black sea Circle Consortium in Educational Research) and ISREV (International Seminar on Religious Education and Values) and ESWTR (European Society of Women in Theological Research). Part of her responsibilities is designing and implementing programmes for in service teacher education; in-service training for school principals and teachers; re-designing and teacher training curricula to include sustainability dimension in teacher training; implementing of action research in teacher training programs, and the preparation of large scale international conferences.

dzintra.ilisko@du.lv

Manisha Avadhut Ketkar is a Commerce graduate from Mumbai University and a Fellow Cost Accountant (FCMA). She has also done her Masters in Business Studies (MBS) from the University of Pune and is currently pursuing her PhD. She has more than 23 years of experience which includes 16 years of industry experience in the area of Operations Management and more than 8 years of teaching experience in Symbiosis Institute of International Business (SIIB) Pune. Her areas of interest are Cost and Management Accounting and Supply Chain Management. She is the Deputy Director at SIIB.

Marcel van der Klink has a position as associate professor at the Welten Institute of the Open University of the Netherlands. Workplace learning, competencies, assessment, professional development and technology-enhanced learning are his main fields of interest. Marcel combines his work at the Open University with a job position as director of the research program ' Educational innovation and teachers professional development at Zuyd, a Dutch university of applied sciences.

Joop de Kraker is associate professor of Environmental and Sustainability Sciences at the Faculty of Management, Science & Technology at the Open Universiteit and at the International Centre for Integrated assessment and Sustainable development at Maastricht University, both in the Netherlands. He is also international coordinator of the European Virtual Seminar on Sustainable Development, an online collaborative learning course jointly organized by 10 European universities, and participated in many EU-projects on e-learning and sustainability. His research interests focus on ways to support joint learning in sustainable development processes. Joop's original background is in agricultural systems ecology and he holds MSc (1989) and PhD (1996) degrees from Wageningen University, the Netherlands.

joop.dekraker@ou.nl

Shilpa Kulkarni is a management graduate from ICFAI university. Currently she is pursuing her interdisciplinary doctoral studies with Symbiosis International University. She is in the role of Assistant professor at the Symbiosis Institute of International Business (SIIB) in the Department of Energy and Environment. Her research interests are corporate sustainability, organizational culture ,sustainability reporting measures.

Andy Lane is Professor of Environmental Systems in the Department of Engineering and Innovation, Faculty of Maths, Computing and Technology at the Open University. Professor Lane was the founding director of the Open University's OpenLearn platform for open education resources *http://openlearn.open.ac.uk/* and has research interests in systems of open education and the management of complex environmental systems. He was Primary Investigator on the SusTEACH research and development project 2011-2013 funded by the UK Jisc Greening ICT programme (*http://www.open.ac.uk/blogs/susteach/*) which was shortlisted for the Green Gown Award 2012 (*http://sustainabilityexchange.ac.uk/research-and-development/ 1377-green-gown-awards-2012-research-a-development-the-open-university-finalist-case-stu dy-a-video*), and later developed as a new teaching unit for the OpenLearn platform (*http://www.open.edu/openlearn/nature-environment/the-environment/the-environmental-im pact-teaching-and-learning/content-section-0*).

Fernando Latorre is Head of the Informatics Department of UNED Barbastro Associate Center. Virtual Attaché's Responsible of the University.

Walter Leal Filho is a professor of environment and technology, teaching at the Hamburg University of Applied Sciences (Germany) and Manchester Metropolitan University (UK). He has over 300 publications on matters related to environment and sustainable development to his credit, and has undertaken a wide range of international projects on sustainability issues. He edited various journals on sustainable development, and coordinated a large network of sustainability experts.

walter.leal@haw-hamburg.de

Rocío Muñoz Mansilla received her Graduate in Physics from Universidad Complutense de Madrid (2000), and the PhD degree in Computer Science and Automation from National Distance University of Spain (UNED). Since 2004 is Assistant Professor of computer science and automatic control at UNED. Her current research interests include system identification and robust control within different research projects. Since 2009, is Vice-Secretary General Technical of UNED.

Ana Paula Teixeira Martinho holds a PhD on Environmental Engineering (2003), a MSc in Sanitary Engineering (1999) and a BSc in Environmental Engineering (1993) from Universidade Nova de Lisboa, Portugal. She is currently Assistant Professor in the Department of Science and Technology at Universidade Aberta and a researcher at the Education Development Research Unit, Portugal. Her main research and teaching areas include environmental impact assessment, integrated waste management, environmental ethics and environmental citizenship and participation, and e-learning in science education. She is now the vice-coordinator of B.Sc in Environmental Sciences.

Paula Bacelar Nicolau, PhD in Environmental Microbiology from University of North Wales, Bangor, UK, is Assistant Professor at Universidade Aberta, Portugal, where she teaches Biology, Microbiology, Conservation Biology, and Food Biochemistry. She is Senior Researcher at the Centre for Functional Ecology of University of Coimbra. Her main research interests include e-learning in biological and environmental sciences, and environmental microbiology. She is the coordinator of the MSc in Environmental Citizenship and Participation at Universidade Aberta.

Paula.Nicolau@uab.pt

Mª Carmen Ortega-Navas was graduated at the Complutense University of Madrid (1988) and received her Doctorate (PhD) in Philosophy and Educational Sciences at the National Distance Education University (UNED) in 2008, Madrid (Spain). Since 2004, she performs her teaching in Health Education, Active and Healthy Aging and Continuing Education and research activities at UNED. She is an Assistant Professor at the Educational Theory and Social Pedagogy at UNED.

Daniel Otto, was educated at the Eberhard Karls Universität Tübingen, Germany, from which he received his Master of Arts in 2008. He is research assistant at the FernUniversität in Hagen, in the Department of Political Science, where he teaches Governance and Interdisciplinary Environmental Science.

Daniel.Otto@FernUni-Hagen.de

Yogesh Patil earned his MSc and PhD in Environmental Sciences from Pune University and currently holds the position of Associate Professor and Head – Research & Publications with Symbiosis Institute of Research and Innovation (SIRI), Symbiosis International University (SIU), Pune, India. He has over 20 years of experience in the field of Environmental Science, Management and Technology. Currently, he has been teaching to post-graduate students in the broad areas of Energy and Environmental Management. His research interest includes developing low-cost integrated technologies/strategies for urban and industrial waste management, climate change and industrial ecology. He has successfully completed six research and consultancy projects funded by national and international agencies like UGC, India; IFS, Sweden and World Bank. He has published over 30 papers in reputed journals and books.

Baltasar Peñate (Ph.D. Chem. Eng.): Head of the Water Department and coordinator of national and international projects on RE powered desalination, non-conventional wastewater treatments, water quality analysis, management and sustainability in water treatment projects. Co-author of the PV-RO patent DESSOL® and the CONTEDES© utility model. Teacher of Renewable Energy Desalination courses. 20 Publications.

Francisca Pérez Salgado is UNESCO-Chair in 'Knowledge Transfer for Sustainable Development supported by ICTs' at the Open University of the Netherlands. She holds a PhD in Physical Chemistry (University of Amsterdam, 1992) and was Dean of the School of Science at the Open University of the Netherlands from 2001 to 2013. Her research focuses on concepts for knowledge construction and on competence and curriculum development in higher education. Societal aspects and impacts play an increasingly important role in natural and environmental sciences. Her research also concerns modern ICT-tools and open education concepts and the way they can improve and assist (e-)learning processes.

paquita.perez@ou.nl

Manuela Pinto, graduated in Educational Sciences, a collaboration scheme led by the portuguese Open University (Universidade Aberta). The Authors of this edition sincerely acknowledge the careful and competente help provided by Ms. Manuela Pinto in the peer review process, communication with the authors and text editing.

Rudi Pretorius is attached to the Department of Geography at the University of South Africa, where he has been lecturing a variety of sustainability related courses over a number of years. In addition he is the academic coordinator of the undergraduate programme in environmental management and currently vice-chair of the department of Geography. He holds a Master's of Science in Geography and a Master's in Business Leadership. He is chair of the Division of Mathematical and Physical Sciences of the South African Academy of Science and Arts and is currently registered for a PhD in Geography with focus on the potential and role of Geography in education for sustainability in the South African context.

Pretorw@unisa.ac.za

Prakash Rao is an Associate Professor and Heads the Energy and Environment Programme at the Symbiosis Institute of International Business, a constituent of Symbiosis International University, Pune, India. He holds a Ph.D. from the University of Bombay and has coordinated several multidisciplinary environment research projects ranging from natural resources to climate change and energy. He has 30 years of experience in the field of sustainable development and environment management with interests in climate change and energy, water, urbanisation sectors and has undertaken assignments for the Government of Qatar, organisations like World Bank,JICA and ICLEI etc, on designing and implementing environmental frameworks including national level policy assessments. He currently teaches post graduate courses on climate change, corporate sustainability, natural resource management at Symbiosis International University and guides doctoral students on various research topics on sustainable development. He has published more than 30 research papers in reputed journals including several articles in magazines and the media and has co authored three books.

prakash.rao@siib.ac.in

Mª Ángeles Rebollo-Catalán, Ph.D., Associate Professor in Methods of Educational Research at the University of Seville (Spain). Her main research area is focused on Gender, Technology and Education, in which she has participated as a researcher of projects funded by several Spanish entities. Her latest articles examine the role of emotions and socially perceived support on instructional processes. She is currently carrying out an excellent project on the impacts of ICT on the well-being and life quality of women.

rebollo@us.es

Anne Sibbel holds post graduate degrees in environmental sciences, education and public health nutrition. As a senior lecturer and principal research fellow at RMIT University, she has worked in the areas of resource management and food sustainability, with a special research interest in exploring ways to promote and embed sustainability within mainstream curricula. She has designed and supervised a number of projects focused on the developing university students' capabilities for sustainability in preparation for the challenges they will face in future professional work and research.

anne.sibbel@rmit.edu.au

Vicente J. Subiela (Mech. Eng.): Head of Section of the Water Department. He has been working on different RE powered systems (solar distillation, wind powered desalination, PV – RO units) since 1998. He has been researcher and coordinator of EU and international cooperation projects, and projects manager of autonomous desalination units. Teacher of Renewable Energy Desalination courses; main creator and coordinator of the e-learning course on "Introduction to desalination by renewable energies". 12 publications.

vsubiela@itccanarias.org

Ed Swithenby was Research Assistant in the Faculty of Maths, Computing and Technology at the Open University on the SusTEACH research and development project on higher education courses and their carbon impacts funded by the UK Jisc Greening ICT programme (*http://www.open.ac.uk/blogs/susteach/*). This was shortlisted for the Green Gown Award 2012 (*http://sustainabilityexchange. ac.uk/research-and-development/1377-green-gown-awards-2012-research-a-development-the-open-university-finalist-case-study-a-video*), and later developed as a new teaching unit for the OpenLearn platform (*http://www.open.edu/open learn/nature-environment/the-environment/the-environmental-impact-teaching-and-learning/content-section-0*).

John Manyitabot Takang is the founding Executive Director of the Cameroon-based Environmental Governance Institute (EGI). He was the Academic Officer of the International Master of Environmental Sciences Programme (IMES) at the University of Cologne for close to four years. He holds a Master of Science in Environmental Sciences from Cologne and a Bachelor in Environmental and Resource Management (ERM) from Cottbus, Germany. He is a PhD Candidate at the University of Cologne and his research focuses on the domestication of multilateral environmental agreements in countries of Sub-Saharan Africa.

john.takang@engov-institute.org

Annie Warren is the Program Director for the Sustainability Science Education Project at Arizona State University. In addition to organizing the workflow and research of the SSE Project, she also helped develop and implement Sustainability Science for Teachers, a course that is required for all preservice elementary teachers within the Teachers College. This hybrid course educates the students through the use of technology, digital storytelling, and real-world explorations, and trains preservice teachers on unique ways of implementing sustainability topics into their future classrooms. Annie is also the Course Coordinator in charge of all course sections, and is a Faculty Associate for the Mary Lou Fulton Teachers College at ASU. Annie holds a BA in Sustainability with a focus on Urban Dynamics from Arizona

State University, as well as a Master's in Science and Technology Policy from ASU's Consortium for Science, Policy, and Outcomes. Additionally, she has a Professional Masters of Interior Architecture from UCLA, and is a LEED Accredited Professional. Currently, she is pursuing her PhD in the Human and Social Dimensions of Science and Technology at ASU.

Annie.warren@asu.edu

Gordon Wilson is Emeritus Professor in Environment and Development at the Open University UK and Visiting Professor at the Open University Netherlands. His research interests concern the processes of knowledge generation and communication for sustainable development, especially the interaction between everyday lived experiences and the sciences. He has published extensively on this and related topics, including two books and 12 papers in scientific journals. From 2009 to 2012 he was coordinator of the European Erasmus Project, 'The lived experience of climate change: e-learning and virtual mobility'.

g.a.wilson@open.ac.uk

Thematic Index

Umweltbildung, Umweltkommunikation und Nachhaltigkeit
Environmental Education, Communication and Sustainability

Herausgegeben von / Edited by Walter Leal Filho

Band 1   Walter Leal Filho (Hrsg.): Umweltschutz und Nachhaltigkeit an Hochschulen. Konzepte-Umsetzung. 1998.

Band 2   Walter Leal Filho / Farrukh Tahir (eds.): Distance Education and Environmental Education. 1998.

Band 3   Walter Leal Filho (ed.): Environmental Engineering: International Perspectives. 1998.

Band 4   Walter Leal Filho / Mauri Ahlberg (eds.): Environmental Education for Sustainability: Good Environment, Good Life. 1998.

Band 5   Walter Leal Filho (ed.): Sustainability and University Life. 1999. 2nd, revised ed. 2000.

Band 6   Wout van den Bor / Peter Holen / Arjen Wals / Walter Leal Filho (eds.): Integration Concepts of Sustainability into Education for Agriculture and Rural Development. 2000.

Band 7   Manfred Oepen / Winfried Hamacher (Eds.): Communicating the Environment. Environmental Communication for Sustainable Development. 2000.

Band 8   Walter Leal Filho (ed.): Communicating Sustainability. 2000.

Band 9   Walter Leal Filho (ed.): Environmental Careers, Environmental Employment and Environmental Training. International Approaches and Contexts. 2001.

Band 10   Rolf Jucker: Our Common Illiteracy. Education as if the Earth and People Mattered. 2002.

Band 11   Walter Leal Filho (ed.): Teaching Sustainability at Universities. Towards curriculum greening. 2002.

Band 12   Walter Leal Filho (ed.): International Experiences on Sustainability. 2002.

Band 13   Lars Wohlers (Hrsg.): Methoden informeller Umweltbildung. 2003.

Band 14   Ulisses Azeiteiro / Fernando Conçalves / Walter Leal Filho / Fernando Morgado / Mário Pereira (eds.): World Trends in Environmental Education. 2004.

Band 15   Walter Leal Filho / Arnolds Ubelis (eds.): Integrative approaches towards sustainability in the Baltic Sea Region. 2004.

Band 16   Walter Leal Filho / Michael Littledyke (eds.): International Perspectives in Environmental Education. 2004.

Band 17   Daniela Krumland: Beitrag der Medien zum politischen Erfolg. Forstwirtschaft und Naturschutz im Politikfeld Wald. 2004.

Band 18   Walter Leal Filho / Bernd Delakowitz (Hrsg.): Umweltmanagement an Hochschulen: Nachhaltigkeitsperspektiven. 2005.

Band 19   Walter Leal Filho / Volker Lüderitz / Gunther Geller (Hrsg.): Perspektiven der Ingenieurökologie in Forschung, Lehre und Praxis. 2006.

Band 20   Walter Leal Filho (ed.): Handbook of Sustainability Research. 2005.

Band 21   Walter Leal Filho / Dieter Greif / Bernd Delakowitz (eds.): Sustainable Chemistry and Biotechnology – A Contribution to Rivers Management. 2006.

Band 22   Walter Leal Filho / David Carpenter (eds.): Sustainability in the Australasian University Context. 2006.

Band 23   Walter Leal Filho / Arnolds Ubelis / Dina Berzina (eds.): Sustainable Development in the Baltic and Beyond. 2006.

Band 24   Walter Leal Filho (ed.): Innovation, Education and Communication for Sustainable Development. 2006.

Band 25   Walter Leal Filho / Mario Salomone (eds.): Innovative Approaches to Education for Sustainable Development. 2006.

Band 26   Walter Leal Filho / Franziska Mannke / Philipp Schmidt-Thomé (eds.): Information, Communication and Education on Climate Change – European Perspectives. 2007.

Band 27   Ulisses Miranda Azeiteiro / Fernando Gonçalves / Ruth Pereira / Mário Jorge Pereira / Walter Leal Filho / Fernando Morgado (eds.): Science and Environmental Education. Towards the Integration of Science Education, Experimental Science Activities and Environmental Education. 2008.

Band 28   Walter Leal Filho / Nils Brandt / Dörte Krahn / Ronald Wennersten (eds.): Conflict Resolution in Coastal Zone Management. 2008.

Band 29   Walter Leal Filho / Franziska Mannke (eds.): Interdisciplinary Aspects of Climate Change. 2009.

Band 30   Mahshid Sotoudeh: Technical Education for Sustainability. An Analysis of Needs in the 21st Century. 2009.

Band 31   Walter Leal Filho (ed.): Sustainability at Universities – Opportunities, Challenges and Trends. 2009.

Band 32   Walter Leal Filho (ed.): World Trends in Education for Sustainable Development. 2011.

Band 33   Fernando Gonçalves / Ruth Pereira / Walter Leal Filho / Ulisses Miranda Azeiteiro (eds.): Contributions to the UN Decade of Education for Sustainable Development. 2012.

Band 34   Walter Leal Filho (ed.): Sustainable Development at Universities: New Horizons. 2012.

Band 35   Ulisses Miranda Azeiteiro / Walter Leal Filho / Sandra Caeiro (eds.): E-Learning and Education for Sustainability. 2014.

www.peterlang.com